T0283379

It's a Gas

It's a Gas

The Sublime and Elusive Elements That Expand Our World

MARK MIODOWNIK

MARINER BOOKS

New York Boston

First published in the United Kingdom by Penguin Books Ltd. 2024.

FIRST US EDITION

Library of Congress Cataloging-in-Publication Data has been applied for.

ISBN 978-0-358-15715-1

24 25 26 27 28 LBC 5 4 3 2 1

For my brothers Sean, Aron and Dan in acknowledgement of our unconventional upbringing and the dreams it inspired.

Contents

Illustrations

Figures

Introduction

I am afraid of the dark. I have been since I was small. Ghosts groan outside my window in the middle of the night and creak the floorboards in the hallway. Science explains this away – branches swaying in the wind, the thermal contraction of wood as it cools – and I believe these rational explanations: I am a scientist. And yet there are still moments when my brain refuses to listen, when irrational fears crowd into my head. Like everyone, there are times – for me, it's being alone in the dark – when my subconscious runs the show, gathering up my anxieties and mixing them with thousands of years of human wonder and superstition about what exists beyond the physical realm.

In the light of day, it seems to me that what gives the subconscious such power – why we *believe* that a ghost might actually be at the end of the bed – is that our irrational brain is often put in charge when the threat is unknown or invisible. We saw this during the COVID-19 pandemic when large numbers of people declared the virus was a hoax and that the television coverage of the millions of people dying around the world was faked. Others declared that the vaccine was a plot to inject an invisible microchip into their bodies to control their minds. This conspiracy theory involved another invisible power: 5G mobile phone signals. The rational majority scoffed at this but carried on harbouring their own irrational beliefs: karma, lucky numbers, horoscopes, gods. I don't think people are stupid for having irrational beliefs, quite the reverse. I think irrationality is an intrinsic part of our human nature. It is a response to the unknown, to living in a complex universe where the agents of cause and effect are hard to detect.

And in truth we are blind to *most* things. Our world is seemingly comprised of stuff we can touch and feel, such as chairs, books and cups of tea. Yet there is a whole class of matter that not only surrounds us but is invisible: gas. Take a breath and you can feel the material reality of air. Its invisibility doesn't make it insignificant though: it has substance – an average person inhales 12 kg of air per day. In outer space it is invisible gas that creates the stars, on Earth it makes our planet habitable. We are enveloped by it, we breathe it, and it saturates us – fumes, vapours, dampness, scents and toxins – all affect our moods, health and behaviour. And yet for most of our lives, we overlook the wonders of gases. Each breath goes unnoticed.

This book is about how the behaviour of gases and the many different types of vapours and whiffs given off by the earth and living organisms enliven, intoxicate and frighten us. These are the gases that first occupied this Earth and have been here ever since. They are ancient and important, and their effects were first interpreted as the work of the gods, of ancient spirits, or the undead.

So what are these substances, really? Physically, they are the stuff we breathe, the stuff that bubbles up in a glass of beer, the stuff that makes flowers smell sweet, the stuff that keeps us alive in hospitals, and yes, the stuff that makes us laugh uncontrollably. Most gases are invisible, odourless and colourless. Even methane, the gas that heats our homes, is undetectable to human senses. The reason you think you can smell it on occasion is because a smelly chemical called methanethiol is added so that leaks can be detected. Over thousands of years, we have pieced together the physics and chemistry that underpin the actions of these gases and given them names like hydrogen, smoke, carbon dioxide, steam, air and hurricanes. Our theories of the atomic nature of gases and their ability to flow, expand and glow explain many of the folk tale phenomena; so does our understanding of how invisible gases can intoxicate, cause hallucinations, and even kill us. They provide some explanation of fairy tales, of ghoulish skulduggery and the power of the dead.

We have also successfully harnessed the power of gases

to create technology that most of us could not live without. It is the precise manipulation of fuel vapour that drives our automobiles; it is the mechanics of steam that still powers our homes; bottled oxygen keeps our loved ones alive; and our ability to remove gases and create a vacuum is the wizardry that helps clean every home. And yet like the sorcerer's apprentice, we have employed these powerful spirits without properly understanding how to control them. For instance, in the 1960s we manufactured chlorofluorocarbon (CFC) gases which had fantastic properties that made them very useful as coolants in refrigerators. It wasn't until the 1970s that scientists noticed a growing hole in the ozone layer in the upper atmosphere and started to blame it on CFC gases. This ozone layer protects life on the planet from high levels of ultraviolet radiation from the sun. Without it we would all be at severe risk of skin cancer. When the ozone hole was first discovered there was denial from sectors of industry selling and using CFCs – their position was that there was no proof that the invisible effects of their invisible products were causing this invisible problem. But the growing ozone hole became such a serious threat to human life that it became politically important to understand what was really going on. Government funding for research followed but it took until 1985 before enough scientific evidence was obtained to determine that it really was CFCs that were causing the hole. A global framework for protection of the ozone layer was agreed, saving the ozone layer for future generations and preventing millions of premature deaths.

Most people don't even know that this invisible disaster

was averted. If the colour of the sky had started changing because of the emission of CFC gases, there would have been uproar at this pollution of our atmosphere. But since we can't see it, people often don't believe it. I include scientists too, for we are not immune to irrationality when it comes to deciphering evidence of cause and effect. For instance, the existence of oxygen was denied by pretty much the whole scientific establishment when it was first proposed as a gas. In the 1750s it was thought that the reason something burns was because it contained a substance called phlogiston. This phlogiston was released when wood burned – the flame was the result of this material escaping. They believed this phlogiston was then absorbed by the air and taken up by plants which in turn explained why they burned – because they had phlogiston in them. This seems very reasonable, doesn't it? It made sense, since it was biological materials that burned while rocks and metals didn't. It wasn't until a scientist called Joseph Priestley discovered a new gas which allowed wood to burn that cracks in the phlogiston theory started to show. Priestley could have rejected the complicated phlogiston theory and pronounced proudly to the world that he had discovered oxygen, a new invisible substance which was the key to explaining flames. Namely, that something burns because it chemically reacts with oxygen gas. But even in the face of his own evidence Priestley stuck to the phlogiston theory, rejecting oxygen. In his lifetime, evidence mounted from others in favour of oxygen as a real gas, and yet Priestley – undeniably one of the most brilliant scientists of his time – was stubborn and irrational about it, believing the phlogiston theory until his death.

We may laugh and shake our heads at Priestley's stubbornness, but there are many modern examples of how scientists hold on to theories about invisible phenomena that are false. For instance, only 150 years ago we were still debating whether helium gas was real, despite it being the second most abundant element in the observable universe. Man-made climate change is the most important example of how invisible cause and effect result in irrational behaviour. The gases that cause global warming are invisible, their accumulation in the atmosphere at more than 400 parts per million is undetectable to human senses. It is easy to deny this kind of thing. Even those who say they believe that man-made climate change is the result of the build-up of carbon dioxide are not acting with urgency. How many people are changing their lifestyles: really buying less stuff, flying less and using less energy? At some very deep level, most of us don't quite believe it enough to act. Carbon dioxide gas might seem to belong solely to the rational world of science, but because it is powerful and invisible, it can feed our delusions and fantasies.

This book is a guide to the gases that inhabit this planet. Along the way we discover that these gases are not just important and interesting, they are our life support system. The emergence of oxygen gas on the planet 2.4 billion years ago radically changed the atmosphere and it literally gave birth to us. Having been created by it, we have learned to manufacture and compress it into tanks that have not just radically improved healthcare but also made uninhabitable places habitable. Oxygen also allows us to explore outer space, the planets and the undersea world. Carbon dioxide, on the other hand, is an often misunderstood part of our life support system. We need carbon dioxide in the atmosphere because without it we would freeze, and all the plants would die. Too much carbon dioxide though and we get global warming, as we are now experiencing. But plant life needs more than carbon dioxide: it also needs nitrogen, which is a plentiful gas in the atmosphere. However, plants can't assimilate nitrogen gas directly from the air, and this limits crop yields. Us clever humans have found a way around this by synthetically harvesting nitrogen from the air and turning it into fertilizer – fertilizer that at a conservative estimate now keeps half of the current world population alive.

We rely on electricity too: it not only keeps us alive by providing energy for heating, cooking and lighting but it is the life support system of our technological world, fuelling our devices and our digital way of life. Power stations create electricity by heating up water using a fuel. That fuel source can be coal, natural gas or nuclear power, but they all turn water into steam, and it is that gas which rotates turbines at high speed to create electricity. Thus steam power is the dominant form of electricity generation on the planet, supplying 70 per cent of it. Over the next century as we move away from fossil fuels, we will be relying on other types of gas technology, such as wind and hydrogen power. Even solar power relies on the precise control of gases and vacuums in its manufacture to ensure the high-purity electronic structures in solar panels are not contaminated. Gas technology powers the world and will continue to do so.

And this reliance on gases affects global economics and politics. This was demonstrated most clearly when Vladimir Putin weaponized the sale of natural gas (methane) as part of the Ukrainian war by turning off the pipelines that connect Russia and Western Europe. In doing so, it caused a political and economic crisis. The price of fertilizer also spiked because methane is the main ingredient in its manufacture, and this in turn increased food prices and caused massive inflation around much of the world. Other gases play an important role in the food system. Nitrogen is the gas inside crisp packets and other dry foods; it is put there to displace air which otherwise would make the contents stale and soggy. Carbon dioxide gas is the most popular additive in drinks, creating the fizz in the sodas, beer and sparkling wines that no one wants to be without. During the COVID pandemic lockdowns there was a shortage in the supply of food-grade carbon dioxide which caused near panic in the UK population, who feared that their favourite afternoon drink, the gin and tonic, would no longer have any fizz. The horror.

Gases play an important part in the fragrance industry too. Perfumes and deodorants are gas technologies as old as history. They are designed to improve our chances of finding love and to influence our mood. As important as perfumes are, cooking and eating are more important, and it is the *smell* of food – gas molecules – that creates most of the experience of flavour. Without a sense of smell, food tastes bland. With that blandness comes a more profound loss, which is the link that smell provides to memory and emotions. It is hard to overemphasize how big an effect smelling something like apple pie baking in the oven can have on our subconscious, bringing to mind memories of childhood and home. Family members who are long dead instantly come alive in our minds at that moment. Sometimes we understand the direct link from a conscious smell to that memory, but mostly we don't. A thought or a memory pops into our head, seemingly unbidden, and yet it is the action of invisible gases on our minds. It is this direct route to our subconscious that marks out gases as such an important part of our emotional life.

In this book I explore how gases became our physiological, technical and emotional life support system: how they breathe life into us, determine our weather and climate, fuel both our technology and our magical beliefs, and make eating so delicious. It is a tour that takes us from the nitrogen that we harvest from the sky to grow our crops to the steam that generates our electricity; from the tanks of oxygen in our hospitals to the tanks of hydrogen that will eventually make flying sustainable; from the pneumatic joy of riding a bicycle to the balm of a summer breeze; from the comedy of farts to the mind-blowing effects of anaesthetic gas. Above all it's a human story of how our relationship with invisible stuff allows us to survive and thrive on this planet.

1. Magical

Driving into central London was an act of extreme stupidity in the 1970s because of the terrible traffic jams and almost complete absence of parking spaces. My mum did it regularly, bundling all four of us boys into the car, with Dad muttering about the madness of it all. In the back we kids were shouting and thumping each other, unconvinced that visiting the National Gallery was going to be fun. My dad meanwhile spent the whole journey doom-mongering about the traffic and the impossibility of finding a parking space. 'Ah shut up,' she would say in her Irish accent, 'it will be fine.' My dad, a scientist devoted to reason, would enter a state of extreme apprehension until my mum drove the car straight up to the gallery entrance just when a car was pulling out of a parking space. My dad would exclaim with utter amazement, 'You've done it again!' and then turn to face us in the back of the car, and say, 'Your mum', shaking his head in admiration for his wife, 'how does she do it?' 'She believes in magic, Dad,' we would say, as if it was as simple as that. He was an unbeliever and therefore destined to never find parking spaces in London.

My mum's reputation as the family witch was cemented in such moments. She believed in all sorts of magical things. Not just ghosts and spirits, but also gods that oversaw the cars and traffic of the metropolis. That machines could have a soul may seem odd in today's world where we see technology as a rational expression of our mastery over nature. But historically the development of autonomous machines has been wrapped up in superstition and the magical arts, and to some it has remained so. Thousands of years ago the Egyptians, the Greeks, the Chinese and then the Arabic cultures used their knowledge of engineering to make many types of machines. Sometimes these were useful things like weapons to protect a city, or pumps to irrigate crops, but alongside them they invented religious statues that moved called automata.

For instance, in Egypt in the third century BCE, during religious festivals a twelve-foot statue of the goddess Nysa could stand, pour milk and sit, without any human intervention. Metal soldiers raised and lowered their spears while guarding Greek halls. Automated golden lions licked water that flanked

entrances to Sultans' palaces in Baghdad. In China a mechanical man made by the artificer Yan Shi not only sang to the
king of the Zhou Dynasty but flirted with the audience,
winking at them. Even though these automata were self-
evidently made by the humans and driven by cogs, weights
and clever mechanisms, these machines were revered as
magical. The engineers and scientists of the time, like Hero
of Alexandria, who created the world's first steam-powered
machine, were thought to have secret knowledge that could
be attributed to their relationship with the gods.

Machines and automata grew in importance in the Middle
Ages, inspiring the development of clocks. The world's first
fully mechanical water clock was designed by Su Song
around 1000 CE. It was ten metres tall and was designed to
mirror the harmony of the universe and so measure the
time it took for the planets to move in the sky. These early
clocks had a spiritual role, by predicting the sun and moon's
movement they connected us to the mystical heavens (this
relationship between mechanical clocks and spirituality is
enshrined today in the location of clock towers on Christian
churches). Over time the mastery of the cog and spring
mechanisms led to more ingenious clockwork automata
with hidden mechanisms such as the mechanical duck
designed by the French engineer Jacques de Vaucanson in
1737. The duck could quack and flap its wings as well as
seeming to be able to eat and poo.

People were wowed by the extreme cleverness of the
machines and started to wonder where the pursuit of such
mechanical power would lead. Would the machines one
day be able to autonomously move, talk and even think? It

was an eighteenth-century innovation by a British Baptist preacher that provided the answer to the first question. His name was Thomas Newcomen and he found a way to make a machine that was driven not by clockwork but by a gas.

The odd thing about steam is its ability to disappear on command and in doing so create an enormous force. You can see this for yourself by carrying out a simple experiment. Take an empty drinks can, put a small amount of water in it and then place it on a stove until the water inside starts to boil (you'll need tongs to hold the can, but it's worth the hassle). This fills the can with steam that displaces the air. Now plunge the can into a basin of cold water. Immediately you witness something extraordinary and magical: the can crushes itself into a tightly mangled lump. It seems to be acted on by an invisible force. Harnessing this force was the key to the development of the steam engine. Others came before Thomas Newcomen to understand steam power, but he usually gets the credit

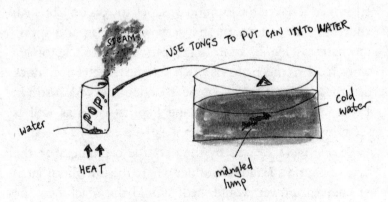

How condensing steam creates large forces

because he was the first person to put it to use profitably. He used it to save the lives of miners.

Mining has always been a dangerous occupation, but mining in seventeenth-century Britain was particularly so. The mines themselves were a collection of narrow, deep, dark tunnels lit by candlelight, and they frequently collapsed, killing those inside. If the miners didn't die from a cave-in they died from asphyxiation by toxic gases, or they died from the fires caused by them, or they died from drowning. By the seventeenth century coal was replacing wood as the primary fuel in Britain and so the number of coal mines was increasing. As the mines became central to the economy of Britain, so these tragedies increased. Floods were the biggest threat to this supply of coal. It rains a lot in Britain, and a mine is a deep hole in the ground, so they tended to fill with water. Pumps were used to remove the water; they were driven by teams of horses who walked in circles harnessed to a rotating drum connected to long ropes or chains that pulled up buckets of water. But this didn't work very well. So, understandably, the mine owners asked the scientists of the day to provide better solutions.

This is where Thomas Newcomen enters the story. He wasn't a scientist but a religious preacher and a blacksmith. But through his practical understanding of machines he had learned the principles underpinning the crushing force caused by cooling steam. When liquid water reaches a boiling point the water molecules have enough energy to become a gas. Liquid water at this temperature bubbles furiously producing this gas, but the boiling point depends on the pressure of the air weighing down on the kettle, the so-called air pressure. We don't normally notice this pressure of the air because it is always there. You can think of the planet being surrounded by an ocean of air. At the bottom of this ocean is where we live on the crust of the Earth. The hundred kilometres of air above us has a weight due to gravity and so it presses down on us – this is called atmospheric pressure.

You don't feel this pressure because you were born at the bottom of this ocean of air and it's always been part of your life. Nevertheless, that pressure is always squeezing you. It is squeezing you now. This pressure also acts on the water in the kettle. As the temperature in the kettle increases the water molecules want to leave the liquid and become a gas, steam. But the atmospheric pressure is squeezing them back into the liquid. The boiling point of water is the temperature at which the energy of the water molecules overpowers the pressure of the atmosphere. At sea level it requires a temperature of 100°C to do this. But if you climb a mountain, there is less air above you, and the atmospheric pressure is lower and so water can boil at a lower temperature. At the top of Mount Everest, where it is almost nine kilometres above sea level, water boils at 71°C.

Understanding atmospheric pressure is the key to understanding the self-crushing can. By heating a small amount of water in the empty drinks can, at first the air inside the drinks can is at atmospheric pressure. This pressure equals the pressure outside the can, and so the sides of the can experience no force. As the steam fills the container it displaces the air creating an equal pressure of steam, and so there is still no force. But now if you plunge the drinks can into water, this immediately cools the steam to create liquid water inside and removing the gas. At this moment the internal gas pressure becomes almost zero. One minute the gas is there pushing against the atmospheric pressure and the next minute it's gone in a puff of non-smoke. A disappearing act. A magic trick. For the drinks can, the pressure of a hundred kilometres of air outside is still very much there. This air pressure is the crushing force. And it's big. Easily able to crush a drinks can.

Newcomen harnessed this power of the atmosphere in his engine. He used a cylinder like a drinks can but with a movable top called a piston, and so when he condensed the steam inside his cylinder by spraying cool water inside, instead of the sides crushing, the piston was crushed down by atmospheric pressure above it. Thus steam in, piston up, steam condensed, piston crushed down, piston up, piston down, piston up, piston down, puff, puff, puff . . . for as long as he had enough coal to burn to create the steam. This was a significant moment in history. It was the birth of the steam engine, and its puff of wispy smoke along with its friendly 'toot' of the steam when directed through a whistle melted the hearts of all engineers who worked with them. The age of steam arrived and brought with it a sense of optimism about what was possible with machines. The life-like automata of ancient times combined with the clockwork cleverness of the Middle Ages now had a powerful spirit in the form of steam that inspired a generation of inventors.

Part of this generation was the Lunar Society, which was founded in the 1760s by engineers, scientists and intellectuals to discuss how to invent the future. They called themselves

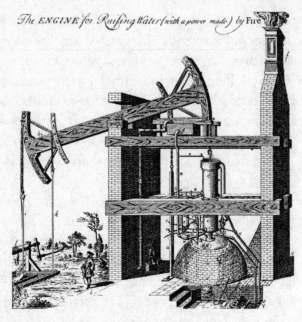

Newcomen's steam engine, 1712

the Lunar Society because they met on evenings with a full moon, which allowed them to ride home safely down the dark country lanes by the light of the moon. They used those occasions to argue and discuss but also to demonstrate to each other their latest discoveries. So when one of the founding members, Erasmus Darwin (the grandfather of Charles Darwin), became obsessed with the idea of creating a machine that could mimic human speech, none of the Lunar Society were surprised. Darwin created several prototypes, which he demonstrated at Lunar Society meetings, although his invention couldn't do much more than mumble, much to the hilarity of the assembled company.

But there was no sense in which this group thought Darwin was crazy for attempting this, even though they referred to themselves as 'lunatics'. Everything was possible, as they saw it. They had a sense of mad optimism about their ability to control the world. One of the Lunar Society members who saw that demonstration, Matthew Boulton, went so far as to offer Darwin £1,000 (£200,000 in today's money) if he could make a robotic machine that could recite the Christian Lord's Prayer, to prove perhaps that such innovations were endorsed by God.

Another member of the Lunar Society, James Watt, became obsessed with the steam engine and decided that it was not good enough. Every time the cylinder of a Newcomen engine was filled with steam it got hot, then water was sprayed in the cylinder which condensed the steam to create the crushing force to bring the piston back down. Spraying water cooled the cylinder, which then had to be heated up again by steam – that was inefficient. If the cylinder could be kept hot, Watt reasoned, then that would save a lot of energy. How to do it is simple once you know the trick. Watt redesigned the engine so that at the top of the piston's stroke, the hot steam was vented out of the cylinder into a separate container and cooled by liquid water. That separate condenser (as it was called) would always be cool and the cylinder would always be hot. Brilliant. Watt, aged thirty-three, had made the steam engine vastly more efficient.

But despite the greater effectiveness of Watt's engines, he didn't know how to make money out of them. Even the business of pumping water out of mines didn't attract customers: the Newcomen steam engines worked reliably day and night and the mine owners saw little incentive to invest in new ones because coal was so cheap – they owned the coal mines, after all. This is where a fellow member of the Lunar Society, Matthew Boulton, comes into the story. He came up with a new business model for selling Watt's steam engines. They offered to replace the Newcomen steam engine with a Watt steam engine in return for a cut of the coal savings made from replacing the inefficient Newcomen engines. The more efficient the replacement engines were, the more customers would benefit and the more money Boulton and Watt would make – it was win–win.

This approach didn't work for those who didn't already have a Newcomen engine. So Boulton and Watt came up with another way to calculate the savings of a steam engine over other ways of powering machinery such as horses to turn a mill wheel to grind flour. Watt calculated that a typical horse could turn a mill wheel 144 times an hour. This was a measure of the power of a horse. He then built steam engines that could match and better this quantity of 'horse power'. This allowed Boulton and Watt to use the same financial model to calculate the efficiency gains of installing one of their steam engines in a mill. The success of this approach meant that the term 'horsepower' became a widely used measure of power. Even now the horsepower of a car is quoted as a measure of its performance. An average compact car for instance has an engine rated at 260 horsepower, which doesn't literally mean that it's equivalent to 260 horses pulling you along, but at the same time it kind of does. The horsepower of an engine is still quoted: the rocket that took Apollo 11 to the Moon was

160,000,000 horsepower (the modern unit of power is called the Watt and is named after James Watt).

In 1764 Erasmus Darwin grabbed James Watt at one of their Lunar Society dinners and mumbled into his ear that he had a much better idea for the steam engine than the tedious business of pumping water. He proposed a new mode of personal transport powered by steam, a steam car. It wasn't just talk: he had designed one that he wanted Watt to see. The apparent madness of his design – a massive Watt engine on top of a cart – doesn't undermine the seriousness of Darwin's proposal. He wrote letters to Matthew Boulton about it too, asking questions about its feasibility and whether Boulton would want to become a business partner in producing these machines – even going so far as to ask how much water and coal Watt's steam engines typically used, so he could calculate how much of these commodities his fiery chariot would need to carry, along with the passengers.

Darwin dreamed of a country where citizens no longer walked miles in the mud but instead jumped into their steam cars and buzzed up and down the lanes, 'toot tooting' each other. But his design in 1764 was not viable: it was too heavy and envisaged a chain mechanism to turn the up–down motion of the steam engine into a rotational motion to turn the wheels. Six years later a Frenchman, Nicolas-Joseph Cugnot, solved that problem by using a ratchet mechanism to create rotational motion and in doing so created the first gas-powered motor vehicle. The car he built weighed about 2.5 tonnes and had the speed of about walking pace. This doesn't sound very useful, but the military were interested in its ability to carry heavy weapons and equipment. And it

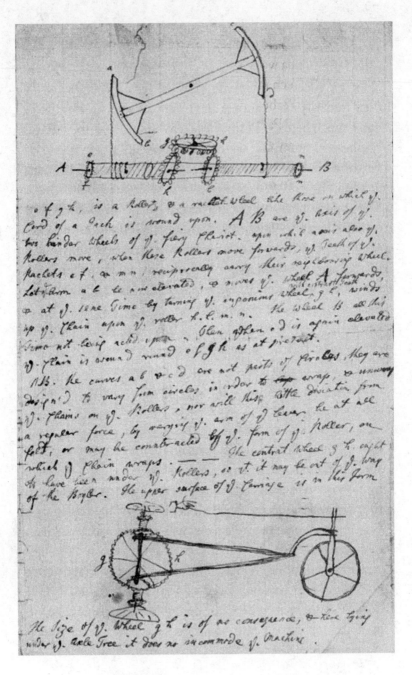

Erasmus Darwin's designs for a steam car, 1764

worked, as testified by records of the time but also because a working replica was built in 2010 by students at the Arts et Métiers Paris Tech (a video of it is available online).

So why didn't the steam car take off in 1770 with France leading the way? Part of the answer is undoubtedly that the steam technology at the time was basic and low power. The vehicle used water to create the steam and had to carry this around with it, so once it ran out of water it ran out of puff. The fuel was coal, and this had to be carried too, adding to the weight. The engine was extremely heavy, and so overall it had a low power-to-weight ratio. This made the vehicle unwieldy, and unless it was on a flat road it was unstable with, according to reports, the coal fire going out every fifteen minutes. It was difficult to steer and frequently went

Cugnot's steam car

out of control and crashed into walls and buildings. In short, the machine couldn't yet compete with horses, which admittedly also needed water and fuel, but needed it less often, and could cope with all terrains.

The steam engines that Boulton and Watt were championing in the eighteenth century were enormously heavy too, but they were static, and so their power-to-weight ratio was not important for the tasks they were doing. Watt knew, as did everyone else in the steam engine world, that you had to use high-pressure steam to increase the power-to-weight ratio. He wasn't mad enough to try it though, because those who did inevitably died.

In a high-pressure steam engine, the pistons are not pushed by atmospheric pressure, as in early versions of the steam engine, but instead by the high-pressure steam. This produces a lot more power without having to increase the size and weight of the boiler. The problem with developing these engines was that high-pressure steam put them under much higher stress, so if they failed, high-pressure steam spurted out, inflicting horrible burns on anyone in the way. Often the boilers exploded and fired shards of metal into anyone nearby. So it took a particularly brave engineer, such as the Cornishman Richard Trevithick, to try enough times, and to fail enough times, to finally succeed whilst not killing himself. It required higher-precision engineering, much better steel, and safety valves. He also needed to make sure his design did not infringe any of Watt's patents because otherwise he would have to pay a licence fee and reduce the profitability of his new steam engine. Thus there was no condenser, but instead the steam was vented up a chimney, creating the 'chuff chuff chuff' so characteristic of steam travel. He demonstrated his steam carriage, which he called the 'Puffin' Devil', on Christmas Eve of 1801 by carrying six passengers up Camborne Hill at approximately 9 mph without killing any of them. His Puffin' Devil did malfunction and destroy itself the next day but that was because they left the boiler fire burning while they were in the pub eating roast goose.

Having proved the viability of high-pressure steam, Trevithick hit an unforeseen problem – that of the quality of the roads. His engines had a much higher power-to-weight ratio than previous steam engines, but his were still very heavy. Without a system of well-made and well-maintained roads the steam carriage would hit potholes, slip and slide in the mud, or destroy the road surface due to its huge weight. The competitor technology was a horse and carriage, and being thousands of years old it was more robust and able to cope with bad roads – it was also considerably lighter, faster and cheaper. If steam-powered cars were ever going to catch on, then they would need a reliable road network able to cope with steam carriages travelling along them at 30 mph.

There was still another way forward for steam travel and that was on water. Boats float on water even when they are made of steel because their hulls displace a larger weight of water than the weight of the ship. Adding a heavy engine wasn't a problem. Big boats need deeper rivers, canals and docks, but there wasn't the issue of the weight of the vehicle destroying the road surface as there was with steam cars. Sea-going steamers in particular could slot straight into the docking infrastructure of sailing boats and had a commercial advantage in that they could sail against the wind or when there was no wind. This was particularly important for navigating large rivers and trade routes. So the age of steam travel began not on the road but on water in the early 1800s.

Ocean liners, especially, transformed travel for passengers. These steam ships were not only faster than sail but they became much bigger. By the middle of the nineteenth century, pioneers like Isambard Kingdom Brunel were building ships of astonishing size and power, like the SS *Great Eastern* capable of carrying 4,000 passengers from

Lounge of steam ocean liner

England to Australia without refuelling. These magnificent steamboats turned into floating luxury hotels with fine dining, ballrooms, thirty-piece orchestras, champagne, running water and electricity for the first-class passengers.

Once the magic and power of steam transport had shown itself to be useful on the water it returned to the land, but this required imagination: it required the steam engineers to think beyond roads for people. George Stephenson had that enthusiasm and vision. Born into a mining family in northern England to illiterate parents, Stephenson was a self-taught engineer and keen to help his fellow workers live a better life. It was he who invented the first safety lantern for mines to prevent explosions due to flammable gases in 1815. He demonstrated it a month before the celebrated Humphry Davy presented his version to the Royal Society – although being from a poor family, having no formal education, and speaking with a strong Northumbrian accent meant that Stephenson's claim was at first dismissed by the UK Parliament. In the end he was credited with an equal claim to the invention with Humphry Davy. The episode had an important impact on George, who made a point of having his son, Robert, privately educated so he could speak with the accent of the British ruling class. Thus when later they built the steam locomotives, it was Robert Stephenson who could represent them in Parliament for the important matters of obtaining regulatory approval for the development of this new form of transport.

The father-and-son team built the first commercial railway locomotive, called 'The Locomotion', capable of hauling eighty tonnes of coal and reaching speeds of 24 mph. Originally the Stephensons conceived rail as a transport system for coal and flour, but they also built 'The Experiment', the first railway carriage for passengers. Thus Stephenson and son became the first to show that rail was the future of steam transport on land, not the car. They did this by not only designing the locomotives, but crucially by designing the whole transport system – the rails, the stations, the bridges and the tunnels. By having control over it all, they could make it work in a way that steam car manufacturers, by not being able to control the roads and signalling systems, could not. The few steam cars that did arrive on British roadways were subject to strict governmental regulations. For instance, in the 1830s, just as the first inter-city railway connecting Liverpool to Manchester was opening (powered by George Stephenson's locomotive 'Rocket'), the UK Parliament passed legislation in the form of the Turnpike Acts, which imposed heavy tolls on steam road vehicles. The rationale was that building and maintaining a road system to support horses and carriages was expensive enough, and that steam cars were too heavy and destroyed the roads. The 1865 Locomotive Act imposed speed limits on steam cars of 2 mph in towns and 4 mph in the countryside. Also, probably as a way to ridicule steam cars, they imposed a law that a person bearing a red flag was legally obliged to walk in front of every vehicle.

Access to transport has always been tied to social equity. Before the time of the steam railways, if you wanted to travel from your home to a town ten or fifty miles away you walked with your luggage if you were poor, and if you were rich, you got a horse-drawn coach, but even then you had a bumpy, muddy, uncomfortable and often unpredictable ride. If you lived inland and wondered what the sea looked like, you probably never saw it in your lifetime, let alone had a beach holiday. The railways changed this, not just for the rich but more importantly for poor people – since they were designed as a mass transport system, the economics only worked if it was affordable to everyone. By 1838 the steam railways in the UK had 5.5 million passenger journeys, by 1845 it was 30 million, and by 1855 it was 111 million. This was nothing short of a revolution in travel, of personal horizons and opportunities for everyone rich and poor alike. No longer were farm labourers stranded in their village for their whole life; they could do more than dream about what was beyond the next hill; they could travel to the big cities, to the beach, to the port to catch a steamer and from there travel the world.

Of course, not everyone was so sure it was a good idea. There were plenty at the time who saw the development of steam railways as a dangerous break with the past: a dirty smoke-polluting industrialization of a bucolic landscape, a desecration of a green and pleasant land. While poets such as Ruskin and Wordsworth campaigned against the railways on these aesthetic grounds, medical doctors warned that they would make people ill and that humans could not survive travelling at speeds greater than a horse: there were invisible forces that would cause harm. Indeed, there were many stories in the newspapers of people going mad after boarding a train, taking all their clothes off for instance, and screaming out of the window.

Other stories extolled the virtues of this new era of travel. Perhaps the most famous was Jules Verne's *Around the World in Eighty Days*, published in 1873. The plot revolves around a bet that Phileas Fogg makes with his friends at the Reform Club in London. He bets them that he can circumnavigate the world in eighty days. It is an adventure tale but not one that features a heroic action figure hacking his way through a jungle, nor is it one about captaining a boat in stormy seas, nor is it about climbing an alpine pass, nor is it about fighting the enemy. Instead the story centres on the urbane gentleman Phileas Fogg, who upon analysing steamer and train timetables calculates that he can theoretically connect the whole world. There are adventures on the way, but the real tension is whether Fogg will lose the bet and his fortune. Could he leave London on a steamer, get to Suez and then cross the

The cathedral-like space of Paddington Station
in the golden age of steam locomotives

Indian Ocean and arrive in Bombay in time to catch a steam train across the whole sub-continent of India to arrive in Calcutta? Did that railway even exist? In the story Fogg consults a Bradshaw's steamer and railway guide, a real series of books that catalogued the timetables of the world's steam transport.

Around the World in Eighty Days was serialized in magazines and some thought Fogg was a real person and that the journey was being reported as he attempted to win his bet. Indeed, such was the excitement about the book that it inspired real attempts, and by 1889 a woman – Nellie Bly – actually did it, beating her fictional rival and travelling around the world in seventy-two days.

Steam trains liberated people, and increasingly women, to travel independently in safety. Not at first of course. At first there were rickety uncomfortable carriages and tragedies such as the boiler explosion that killed two people in Carlisle station in 1844. But that did not put people off – they wanted to travel too much. As people moved off the land and into the cities, the freedom of steam travel became part of the deal. Steam trains made a different life possible, seeing the world in comfort and style was a liberating opportunity. The architecture of terminus stations in all the capitals of the world reflected that ambition and power. They resembled vast cathedrals of steel and glass to glorify the spirit of steam.

By the end of the nineteenth century, the railways spanned the globe and changed the horizons of practically everyone, and yet despite all this progress Erasmus Darwin's dream still had not been realized: there were no steam cars. Small lightweight steam cars that didn't damage the roads did come into existence at the beginning of the twentieth century, but they were not popular. It took the invention of the internal combustion engine to bring Darwin's 200-year-old dream to a mass-market reality. These engines didn't use steam but a flammable gas to push pistons up and down in engines – these gases are the vapours of petrol and diesel fuels.

Despite being replaced by petrol, diesel and now electric engines, steam power is still a core part of our modern world. We rarely see steam engines, hear their friendly 'toot', or smell the coal smoke from their boilers, but steam is used to generate almost all our electricity. It is generated in giant boilers in power stations using either coal, oil, gas or nuclear energy. That steam is fed into steam turbines that rotate electric coils at high speeds to create the electricity that powers hospitals, homes, cities, electric trains, electric cars, street lights and everything with a plug. Thus steam gas is still a vital part of our life support system, currently accounting for approximately 70 per cent of the world's electricity. 'Running out of steam', a phrase from our industrial past, would still bring our modern world to a halt.

While in my childhood I may have been deeply sceptical of my mum's belief in the gods of parking, it did seem undeniable that she had some affinity with the world of machines. Her belief reaped dividends for us as a family – she really could get a car parking space anywhere. My own devoutness was directed at sweet things: I worshipped whipped cream. And as we will see next, the gas that delivers this delicious fluffiness is, like steam, another technological spirit at the heart of our life support system.

2. Delusional

Playing rugby one day I mistimed a tackle and dislocated one of my fingers. I felt a stab of pain and sat in the mud staring in disbelief at my hand. One of my fingers was now bent at an unnatural angle. My teammates gathered round grinning. They made it clear that they expected me to snap my finger back into place. 'Go on! Don't be a wimp!' Several of them mimed the action and excitedly made the cracking sound they wanted to hear. This was a test of my toughness, one of many that I have failed in my life. Unlike them I saw myself as mentally tough but not physically brave. More Sherlock Holmes than Rambo, I explained to myself as I trudged off to hospital.

I sat for hours in a blaze of fluorescent lights in the Accident and Emergency Department, feeling self-conscious with my shorts, muddy legs and, of course, my odd-looking finger. Later that evening I was shown into a booth and blue paper curtains were drawn around me for privacy. Presently a young male doctor came in. He seemed distracted but asked a few questions while consulting a clipboard containing my patient info. 'Dislocated finger?' he said, and I nodded. He put the clipboard down, took the dislocated finger in his hand and then with no warning he yanked it violently. There is an absurd moment from this scene that is freeze-framed in my memory for ever. I am screaming in pain and his face is very near to mine. He is red from the effort of trying to pull my finger back into its socket, but he has failed, and instead has managed to pull me out of my seat and right up close to him. We both fall back towards the wall like two drunks fighting.

Afterwards, calming me down with a cup of tea, he explained that he was sorry and that he had been trying to catch me off-guard in order to snap the finger back into place. A manoeuvre that had always worked in the past, he said. Because he was so embarrassed and also perhaps because he was worried that I might complain to the hospital management, he quickly fetched a canister of laughing gas. Attached to the metal tank were a tube and a mask, which he fitted over my nose and mouth. Fiddling with the valves on the canister he asked me to take a deep breath. I did so and felt nothing. The doctor fiddled with the valves some more and asked me to have another go. It felt good, very good. I took another deep breath and soon found myself out of my mind and on a golf course, unable to find my ball.

As with the advent of steam power, the story of how a gas that creates such delusions became a method of standard pain relief begins in industrial Britain. The air in cities and towns was full of coal smoke and often foul-smelling because of the large number of animals and humans in close proximity with little sewage management. It was widely believed that bad air could cause disease. This was the miasma theory. It seemed like common sense because where bad smells were most concentrated, like in cities, there was the most disease. Outbreaks of cholera, bubonic plague and other contagious diseases were thought to be caused by clouds of miasma carried in the stench. If air could carry diseases, then perhaps gas could cure them too – so went the reasoning of Thomas Beddoes, who set up the Pneumatic Institution in 1799 in the city of Bristol.

People died in their millions of respiratory diseases such as consumption (tuberculosis). This is a wasting disease in which the patient becomes weaker and weaker as the infection 'consumes' the body. Finally, after months – or years – of coughing up blood, they die painfully in their bed. For thousands of years, it affected rich and poor alike, although because it was an airborne infection, the poor who lived in crowded conditions contracted it more often. Parents who had consumptive sons and daughters were desperate to find cures, one of which was called Alderney Cow Therapy, which involved inhaling the breath of cows.

This seems mad in the light of what we currently know about the bacterial cause of tuberculosis but back then it seemed reasonable to try anything that might work if it would save your daughter, or son, or wife, or husband – the disease was indiscriminate.

The idea was to find specific gases that could cure specific diseases. Beddoes hired a young man called Humphry Davy, who was only eighteen years old, to carry out the research in gas therapy. Davy tried many gases, none of which seemed to be much help – in fact quite the opposite: when he tried the new gas carbon monoxide on himself he almost died. Undeterred in his zeal to make great discoveries and help humankind he then inhaled another newly discovered gas, nitrous oxide. The gas tasted slightly sweet and had a very strange effect on him – he started dancing round his laboratory 'like a madman', as he noted later. He laughed. He giggled. It was highly inappropriate given that he was in charge of a medical institute, but he couldn't stop.

Thankfully, the effects did wear off after an hour. But how miraculous, how strange, he reflected. He tried it again, sometimes with the same ludicrous effects, and sometimes he was taken out of his mind in a more transcendental way. He wrote 'nothing exists but thoughts' after one session. Davy described breathing nitrous oxide as a sublime experience, an experience that was beyond language, although he did try to capture the feeling in his poetry, describing his limbs as 'clad with new-born mightiness'. Given the plight of his patients and their illnesses, Davy thought he would try the gas out on them to assess if it had any therapeutic value for curing diseases such as consumption. It didn't. But it did make them laugh. He himself found it so delightful that he could not resist carrying on experimenting with nitrous oxide at night. Davy found that breathing the gas heightened his senses – he felt he could see and hear more vividly. He invited friends, writers and poets, such as Samuel Taylor Coleridge, to his laboratory to inhale the gas. They breathed it in and fell into fits of laughter too, loudly singing and dancing. The poet Robert Southey grandly declared that Davy had invented a new type of pleasure which gave 'delightful sensation in every limb – in every part of the body – to the very teeth'. More and more people came to try this remarkable laughing gas, including the radical poet Anna Barbauld. These laughing gas gatherings gave the Pneumatic Institution a revolutionary air. It was the cool place to be.

Remarkably, during all the partying and dancing, Humphry Davy still had the acumen to deduce something scientifically important about laughing gas: it was not just funny and distracting – it could eliminate pain. Up until this point, for pretty much the whole of history, surgery and dentistry had been largely carried out without anaesthetic. For most people this meant living with the agony of toothache until they could stand the pain no more. They would then reluctantly go to a dentist or doctor who would remove the tooth by kneeling on their chest and tugging it out with a pair of pliers. Similarly, those with the excruciating pain of gallstones (small stones that form in the gall bladder, an organ near the liver) often preferred to live with the pain rather than go for surgery to remove them. Alcohol and herbal concoctions containing opium and henbane (psychoactive substances derived from plants) were offered as sedatives, but patients still felt excruciating pain and writhed in agony because these substances didn't effectively block pain receptors. To carry out surgery or amputations, the patients were tied down or held down by others, with a piece of wood or leather placed in their mouth to stop them screaming.

There was also a belief held by Western surgeons that pain itself might be important to the success of the surgery – they thought that it might be required for nature's healing powers to be triggered. Thus, strangely, there was no obvious demand from medical doctors for the development of anaesthetics. So although Humphry Davy discovered nitrous oxide to be a fast-acting anaesthetic that blocked pain receptors, the medical profession wasn't interested.

There were other reasons why Davy didn't develop the anaesthetic applications of laughing gas. Despite his open spirit of inquiry and genuine desire to help humanity, his laughing gas parties eventually gained the Pneumatic Institution a reputation as a place of strange radical experiments. It was a place where the mind-altering drugs were administered to both men and women, causing rumours about inappropriate behaviour. Thomas Beddoes and Humphry Davy were dubbed the Pneumatic Revellers and mocked in the newspapers, which suggested that gas-induced sex parties were taking place under the guise of lofty poetic and scientific experimentation. There was an accusation that a female patient had been made pregnant while out of

Davy was lampooned in the press for his promotion of laughing gas, 1830

her mind on the gas. Others called them pompous 'aeolists', who pretended they had found a secret gateway to a sublime spiritual world. Davy, who wanted to be a serious scientist, left the Pneumatic Institute and moved to the Royal Institution in London to develop his revolutionary work on electricity, and ceased working on the therapeutic applications of laughing gas.

Others were not so worried about the press, especially if the attention could make them money. Samuel Colt was one such person: he wanted to develop a new type of gun but needed money to fund it. In 1832 he decided to tour the United States of America performing laughing gas demonstrations on stage. Colt was a self-taught engineer, and it wasn't hard for him to learn how to make the gas. The formula for laughing gas is N_2O, which means it is made of two nitrogen atoms and one oxygen atom. Since the air we breathe is mostly made up of nitrogen (78 per cent) and oxygen (21 per cent) you might expect that laughing gas would occur naturally. But although the oxygen in the air is very reactive, the nitrogen is not. It occurs as a molecule, N_2, which is to say two nitrogen atoms chemically bound together into a single molecule. This molecule is very stable and reacts with very few things – not even the oxygen in air. The method Samuel Colt used to produce nitrous oxide was to heat up ammonium nitrate, a fertilizer which decomposes to produce N_2O gas. But he needed to be careful. Heating it too fast causes a different reaction, creating enormous amounts of nitrogen and nitrogen dioxide gas very quickly. When large amounts of gas are produced it has to go somewhere, and so it expands outwards. This creates a pressure wave destroying objects in its path and carrying the smashed pieces along with it. In other words, it creates an explosion. This is how dynamite and TNT work. Explosions are chain reactions that produce large amounts of gas. For instance, the Lebanese port of Beirut was flattened in 2020 when a store containing 2,750 tonnes of ammonium nitrate fertilizer heated up as a result of a fire and exploded.

To avoid blowing himself up Samuel Colt carefully heated ammonium nitrate, keeping the temperature below 300°C, and collected the gas that was created in a fine silk bag, which gradually expanded into a balloon. His stage shows involved inviting volunteers onto the stage to inhale

A GRAND
EXHIBITION

OF THE EFFECTS PRODUCED BY INHALING
NITROUS OXIDE, EXHILERATING, OR
LAUGHING GAS!

WILL BE GIVEN AT *The Mason Hall* *interday* EVENING, *15*

50 GALLONS OF GAS

will be
prepared and administered
to all in the audience
who desire to inhale it.

MEN will be invited from the audience, to protect those under the influence of the Gas from injuring themselves or others. This course is adopted that no apprehension of danger may be entertained. Probably no one will attempt to fight.

THE EFFECT OF THE GAS is to make those who inhale it, either

LAUGH, SING, DANCE, SPEAK OR FIGHT, &c. &c.

according to the leading trait of their character. They seem to retain consciousness enough not to say or do that which they would have occasion to regret.

N. B. The Gas will be administered only to gentlemen of the first respectability. The object is to make the entertainment in every respect, a genteel affair.

Those who inhale the Gas once, are always anxious to inhale it the second time. There is not an exception to this rule.

No language can describe the delightful sensation produced. Robert Southey, (poet) once said that "the atmosphere of the highest of all possible heavens must be composed of this Gas."

For a full account of the effect produced upon some of the most distinguished men of Europe, see Hooper's Medical Dictionary, under the head of Nitrogen.

Date: 1845.   #409, Buck Hill Associates, Johnsburg, N.Y.

Nitrous oxide entertainments were popular in the nineteenth century; this poster is from the Museum of the City of New York, 1846 the gas, whereupon they would fall into hysterics, sing and dance. The spectacle of a prim middle-aged nurse suddenly bursting into song, or a shy gentleman transforming into a comedian, or a plain-speaking farmer becoming a poet, provided the entertainment to the paying audience.

Keeping the spectacle in the realms of good taste for the American audience was a priority given the dubious reputation of laughing gas, and so Colt pretended to be a doctor to exert moral authority and create an atmosphere of family entertainment. It wasn't easy and Colt stopped doing the shows once he had raised enough cash for his real passion: the development of a new type of pistol with a rotating cylinder that could fire multiple times without needing to be reloaded. When he perfected it, he named it the Colt revolver. It is the type of gun featured in every western movie, holstered on a leather belt slung off the waist of every cowboy. The revolver was also the gun favoured by Sherlock Holmes in his confrontations with the criminal fraternity. He is a character who enters this story shortly.

After Colt it was a dentist called Horace Wells who got the laughing bug. In 1844 he attended a laughing gas show and wondered whether it might work as pain relief for teeth extraction. He tried it on himself while having a wisdom tooth extracted. He giggled while spitting blood and realized this was for real: laughing gas really did block pain. After trying it on more than ten other patients he decided to go public and performed a tooth extraction in Massachusetts General Hospital. Unfortunately for Horace Wells the patient let out a small cry during the procedure, and although

Ether and ethanol: they have the same chemical composition
but different molecular structures. One is an anaesthetic
and the other is the alcohol in our drinks

afterwards the patient said he felt very little pain – the 'Ah' he emitted was probably the same 'Ah' that everyone expresses when their mouth is being prised open by a dentist – the conservative medical establishment who had been in attendance pounced on this as proof of frivolous fairground trickery. They dismissed nitrous oxide and ridiculed Wells. Tragically, Wells committed suicide a few years later, partly due to a sense of shame and failure, but he had ignited an interest in anaesthesia. Two years after his demonstration, another dentist, William Morton, used a different substance to anaesthetize a patient in the same hospital. The gas was a vapour of a mysterious liquid called ether.

Ether is a clear but very powerful liquid. If you sniff a bottle of ether you will immediately feel woozy. Like ethanol, the alcohol in drinks such as beer and wine, it is made of two carbon atoms, six hydrogen atoms and one oxygen. But the two molecules differ in the way the atoms are bonded together, which has a big impact on their properties. For instance, their boiling points, the temperatures at which they change from a liquid into a gas, are very different. The boiling point of ether is 35°C, while the boiling point of ethanol is 78°C.

Now, 35°C is not a high temperature: a glass of ether will boil in front of you on a hot summer's day. Although a bottle of pure alcohol will not boil on that same summer's day, you will still be able to smell it, and that means it is still releasing some of its molecules into the air. In fact, as with all liquids, molecules jump into the air and become a gas even before they reach the boiling point: this is called the vapour of a liquid and is why you can smell liquids – it is the bouquet of wine, the warming aroma from a bowl of soup, a waft of perfume. The lower the boiling point compared to room temperature, the more vapour you get from it and generally the smellier it is: ether is very smelly, and no one would describe it as having a pleasant bouquet. It is not dissimilar to the smell of diesel and petrol – to which it is chemically related – and it will knock you out.

If you breathe in ether vapour it goes straight to your lungs, where it infiltrates into the bloodstream and causes rapid intoxication in a matter of minutes. As with alcohol, the effect can be pleasant, but there are several differences that make the gas useful for anaesthesia. When surgeons gave alcohol to patients being cut open, patients still felt pain, often babbling incoherently and striking out with arms flailing around. Ether's molecular structure means that it affects different pathways in the brain, causing a patient to lose consciousness rapidly and become insensitive to the pain of being cut open. Crucially they lie still.

Because the boiling point of ether is so low, it can be easily delivered by getting a patient to breathe in the vapour using a vial of the warmed liquid. The trick of course is to get the dose of vapour inhalation right. Too much and the patient is poisoned, causing severe side-effects such as breathing difficulties and heart rate abnormalities. Too little and the patient will wake up while still cut open. The margin between overdose and underdose is small with ether and this makes it a difficult anaesthetic to control.

Beyond this, the use of ether as an anaesthetic was effective but not ideal. For a start, like alcohol, it is addictive. So once there was general acceptance of the idea that the benefits of anaesthetics might outweigh the risks, others started to explore the vapours of different substances looking for better anaesthetics. One of these experimenters was Dr John Simpson, a Scottish obstetrician living in Edinburgh. Every evening he and two assistants would gather in his sitting room and sniff chemicals to assess their anaesthetic effect. This was a risky thing to do because the toxicity of these chemicals was completely unknown. On a dark November evening of 1847 they gathered to inhale a sweet-smelling volatile liquid called chloroform. At first they found it pleasant and were put 'into a good humour', but then suddenly they all collapsed. When they regained consciousness the next morning Dr Simpson was elated: had he found a new anaesthetic? Not realizing he could have killed himself if he had got the dose wrong, he next tried it on his niece, who felt happy, began singing 'I am an angel' and then dropped to the floor unconscious. She survived though, and deeming it safe Dr Simpson then went on to use chloroform in his medical practice, successfully administering it to mothers in labour to ease their pain. It was a miracle he didn't kill some of them by getting the dose wrong.

Other doctors were outraged, but not about the safety issues. They argued that alleviating the pain of childbirth was morally wrong and that God had ordained that women should suffer while giving birth. Others objected that using a substance to reduce the pain of labour would prevent a baby bonding with their mother. The people who argued these things were mostly men. The moral issues around pain relief became a topic of hot public debate until Queen Victoria in 1853 was administered chloroform while giving birth to Prince Leopold, after which it received the royal stamp of approval. The Queen is later quoted as saying to a friend who also used the pain relief: 'very glad to hear Minnie is going on so well & had the inestimable blessing of chloroform w. no one can ever be sufficiently grateful for'.

The emergence of chloroform ignited the imagination of writers, especially of detective fiction. Arthur Conan Doyle, a doctor himself, was one of the first to include it as a plot device to silently incapacitate victims in his Sherlock Holmes stories. Typically, a villain surprises an innocent victim from behind, perhaps as they walk into a room, and puts a cloth soaked in chloroform over their mouth and nose. The victim sags and collapses to the floor. Holmes himself is portrayed as a Dr Simpson figure who experiments in his lodgings using chemicals, test tubes and microscopes – even experimenting on himself, and he is sometimes found in a state of collapse by his faithful sidekick, Watson. His scientific and forensic approach, along with his encyclopaedic experience of exotic poisons and hallucinatory drugs, is a key feature of his detective genius. Like Humphry Davy, he uses drugs to enter his subconscious, to reveal things about the world that cannot be seen by others.

These detective stories and the increasing ease of access to chloroform led to copycat crimes. Many of these were unsuccessful, since the criminals had underestimated how long it takes for a cloth soaked in chloroform and clamped over a victim's nose and mouth to induce unconsciousness. In reality it takes several minutes of breathing in chloroform to knock someone out. During that time the victim often kicked and scratched their way to freedom. However, this was not widely known. The myth of its instant effect was not helped by the popularity of detective fiction in which chloroform went on being the criminal's knock-out chemical of choice well into the era of cinema at the turn of the twentieth century. By then chloroform's use as a medical anaesthetic was being phased out. It was too toxic (damaging the liver and kidneys in high doses) but also caused unexpected deaths in surgery by interfering with the rhythm of the heart – the so-called 'sudden sniffing death' – which is a dangerous side-effect associated with inhaling many solvents.

This balance between effective pain relief and the risk of side-effects brought laughing gas back into the picture. No, it didn't knock you out, but yes, it did allow you to experience less pain with fewer side-effects. However, it was unregulated and the means of administration – a silk or rubber balloon – was not ideal for dentists or medics because balloons are bulky and over time leak gas into the room where they are stored, which in the case of nitrous oxide, makes everyone giddy. One person who took up the challenge to solve this engineering problem was George Poe, a cousin of the poet and master of the macabre, Edgar Allan Poe. He created a factory in New Jersey in the USA for the mass manufacturing of nitrous oxide in a liquid form, which he then sold in canisters.

This was a clever choice. Liquids are denser than gases, and so a lot of nitrous oxide could be crammed into a small cylinder. The boiling point of nitrous oxide is –88°C, which means that liquid nitrous oxide immediately boils at room temperature. However, pressurizing a gas increases its boiling point, allowing it to be kept as a liquid in a pressurized container (such as gas bottles used to store butane for camping). Opening the valve of the canister released the pressure, instantly transforming the liquid into vapour for use in dentists' practices and hospitals. By 1883 George Poe was supplying 5,000 dentists with medical-grade anaesthetic nitrous oxide in canisters.

George Poe was passionate about the power of gases to help people in pain. He patented a respirator as a safe and systematic way of administering gas to a patient. It consisted of brass cylinders which were fed by a canister of gas that was pumped into a face mask placed over the patient's nose and mouth. With this procedure, he claimed, it was possible to use pure oxygen to bring people back from the dead. To demonstrate this he first experimented on rabbits near his factory – capturing them, asphyxiating them (using chloroform) until their heart stopped, and then reviving them using his respirator. After a lot of dead rabbits, he managed to make the procedure work and then moved on to bigger mammals, like dogs. He would demonstrate this in front of a live audience, asking for a stray dog to be captured from the street before asphyxiating it with chloroform until it stopped twitching and breathing. He would then bring the dog back to life using his respirator. These days we would see such cruel public demonstrations as unacceptable, but, as with the story of nitrous oxide, people found them not just convincing but also entertaining.

Once gases could be compressed cheaply and put in convenient cylinders they were put to all sorts of unexpected uses in the twentieth century. For instance, if you add cream to a nitrous oxide gas cylinder some of the compressed gas dissolves in the fat. If you then open the

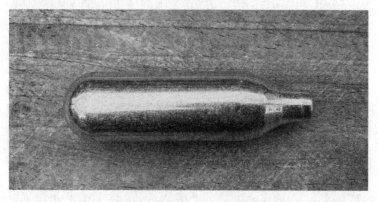

Nitrous oxide whippets used in commercial
kitchens for making whipped cream

cylinder the gas pressure squirts the cream out of the dispenser. As it does this the gas inside the fat expands rapidly and blows billions and trillions of little bubbles – this instantly whips the cream into a fluffy foam. This delicious messy process was discovered accidentally by a chemistry student in the 1930s who was studying the preservation of dairy cream using compressed gas. He tried other gases too, but nitrous oxide works best for cream because, unlike carbon dioxide or oxygen, it doesn't affect the taste. It is also easy and safe to compress into a convenient form called a laughing gas whippet. These look like little bullets that you might load a revolver with, but instead they are designed to fit into a whipped cream gun. Press the trigger and out comes instant and perfect whipped cream: it's delightful, it's delicious, it's a kind of magic.

It is one of the odd things about modern life that something as trivial as whipped cream could spawn an industry and global supply chain whose sole purpose is to deliver boxes of tiny bullet-like whippets of compressed nitrous oxide gas to kitchens all round the world. As a kid, I was of the opinion that there was no dessert in the world that was not improved with whipped cream dispensed in a fluffy pile on top. I still think that now.

The nitrous oxide delivered to hospitals is not in the form of whippets but cylindrical steel bottles. It was one of those that was used by my doctor on the day I dislocated my finger. I breathed in the gas using a respirator mask pioneered by George Poe. These days it is delivered as a mixture of oxygen and nitrous oxide called 'Gas and Air'. Women who give birth in hospital are very likely to be offered this mixture to relieve the pain. It is a simple and easy-to-use system and a much milder painkiller than the epidurals that are the alternative when a woman in labour is in extreme agony. This gas really is part of the life support system of modern hospitals, and indeed our society, delivering pain relief to those who need it.

Whether delivered for pain relief or fluffy cream on desserts, we see this nitrous oxide as a commodity, something squirted out by an industrial system for our use. Laughing gas has lost its role as the poetic muse with a licence to inspire rapture that Humphry Davy first identified in 1799. Rebellious modern teenagers are an exception to this though. In their quest for different ways to inhabit their minds, they have rediscovered laughing gas. They get hold of whippets and use them to blow up balloons of nitrous oxide. They then inhale the gas through the necks of the balloons as a recreational drug. I sometimes find tangible evidence of their hilarious, mad, dangerous (illegal in the UK) and life-changing trips. These are clusters of bright silver whippets that shine like supernatural poo in the green grass of our local park.

My own nitrous oxide trip occurred when I failed to self-regulate the dose I inhaled that day in hospital during my dislocated-finger episode. I floated completely out of my mind. Hearing a 'clack' sound I mistook it for the sound of a ball being hit and became puzzled as to why someone was playing golf in the hospital. Returning to consciousness a few seconds (or minutes?) later, I saw the doctor standing in front of me but there was no sign of his golf clubs. Instead my finger was back where it should be. I had felt no pain this time. The doctor looked pleased.

On the way home from hospital, and despite my injury, I felt elated on the top deck of a bus. It was a dark night and London's grimy streets raced by in the orange glow of the sodium street lamps. Every now and again the bus passed a room on the upper floor of a house in which the occupant had not closed their curtains. I saw snapshots of electrically lit people in their rooms, some sitting in bed reading a book or working on a computer. Then I saw a woman staring out of her window and we locked stares for an instant. I felt like I was inside her head, momentarily transported into her consciousness looking out of her window and seeing a young man on the top deck of a double-decker bus staring at me. Later I rationally attributed this out-of-body experience to the residual effects of laughing gas inhalation. My subconscious mind hangs on to this memory, returning to it frequently, feeling its strangeness like a tongue exploring the hole left after a dental extraction. Once you have experienced moments like this, it is impossible to let go of them, and to dismiss the idea that you have accessed a different realm of consciousness. Of course, there are dangers too, as the next gas shows: the phenomenon can become a route for the supernatural to take hold of the mind.

3. Spooky

The evening of 31 October is the traditional day in the northern hemisphere to celebrate the dead coming back to life. This reflects an ancient belief that as the sun sets, the souls and spirits of an afterworld can seep into ours and meet their loved ones again, or seek redress or revenge. Their presence is felt in the dead of night, in the creak of the stairs of an old house, in the bony caress of your hair by a tree branch. These are not modern feelings – our ancestors have felt them for tens of thousands of years.

We have developed many ways to try to banish these

demons from our psyche, through religion or through superstition and the development of amulets – charms, often worn as necklaces, that depict protective gods or symbols of power. These are still worn by many people today, but the most successful method has been an engineering approach: the development of street lighting. This is the story of a dangerous gas that banished evil spirits from our streets but also came into our homes, and in doing so exacted a price from us. This is the story of methane.

It almost didn't happen. Piping gas throughout a town and into homes by means of a series of tubes and pipes doesn't seem an inevitable route to civilization. It is costly, technically difficult and dangerous. Even today a gas leak must be treated with the utmost respect if explosions and tragedy are not to follow. So why take the risk? The answer lies in the urgent need to banish the evil spirits always seeming to lurk in the darkness. As with the gases we have already met, steam and nitrous oxide, their use is not without risk to human life. Yet in our pursuit of modernity, to travel the world and create powerful machines, to be free from pain during surgery and dentistry, we have had to balance progress with the risk of death, explosions and damage to the environment. And death and destruction did occur, yet we stuck to the development of these gases, creating a modern life support system based on our control of them.

We know that our ancestors lit fires to produce light so they could see in their cave dwellings. The flickering firelight on the walls of the cave enabled them to make tools, clothes and art and to prepare food within a secure environment. It didn't much help them to move around at night though. If you have ever picked up a lighted stick, you'll know how hard it is to keep it alight for any length of time. You need a piece of cloth soaked in a flammable substance such as wax or fat to make a decent torch. Such was its importance that this type of lighted torch became a symbol of hope and security. Lit torches outside notable buildings signified their importance and they were adopted by many civilizations, such as the Greek and Roman empires, to light the way at night. They offered a salvation from the forces of darkness. Religious leaders are often referred to as the light to guide those in the darkness through their lives.

But not all lights were considered good – some were thought to be evil and mischievous. The will-o'-the-wisp was one of these. A ball of flickering light, it appears in woods and marshlands at night. A traveller lost in the woods spotting one of these might think it evidence of a house or small town in the distance and use the light to guide their way to help. But will-o'-the-wisps fall away and flicker as you approach them, and then in the pitch dark leave the traveller lost and deep in the woods. Folklore attributes these false lights to malign spirits in the woods.

In 1667 John Milton, in *Paradise Lost*, attributes this behaviour to evil spirits:

> [As] a flame,
> Which oft, they say, some evil Spirit attends
> Hovering and blazing with delusive light,

Misleads th' amazed night-wanderer from his way
To bogs and mires, and oft through pond or pool,
There swallowed up and lost, from succour far.

(9.637–42)

Every culture has a different name and role for will-o'-the-wisps but the description of the lights is remarkably similar – hovering balls of light that move around marshland or woods at night, luring the foolish to their deaths. In America they are called jack-o'-lanterns (where they are now re-created by candles in carved pumpkins during Halloween); in India they are the ghosts called *chir batti*; in Japan they are lost souls called *hitodama*; in Mexico and South America they are *brujas*, the spirits of witches.

Although the stories associated with will-o'-the-wisps vary across the world – for instance, in Scandinavian countries they are not evil but mark where fairy treasure is buried – the description of a ball of flickering fire is surprisingly consistent. At the time when John Milton was writing about them as evil spirits, scientists were trying to understand if there could be a natural explanation. Marshes were known to bubble occasionally with foul-smelling gas. But the ball-like nature of it was a puzzle. Why would ignited gas stick together as an entity? But then there was another problem, which is that those who had come near to these will-o'-the-wisps reported that they always retreated out of reach while hovering above the ground.

The idea of hiding out in the woods in the dead of night to investigate the physical origin of these evil spirits intimidated most people, but not Major Blesson. In 1811 he was an engineer and major in Napoleon's army and wanted to get to the bottom of what these things could be. From his writings he seems to be a fearless Colonel Deadshot figure who had led troops across Europe and wasn't frightened of anything. He reports seeing the ghostly will-o'-the-wisps at night in the Forest of Gorbitz. Describing the marshy land as having iron-coloured water covered with a strange bubbling crust, he marked the place on a map and plotted a route back to it that he could follow in the dead of night. When he got there, he reported:

> to my great joy I actually observed bluish-purple flames, and did not hesitate to approach them. On my reaching the spot they retired, and I pursued them in vain; all attempts to examine them closely were ineffectual.

He returned another time and found that the will-o'-the-wisps again receded as he approached, but thinking that this might be due to air currents that he had stirred up, he stayed motionless for a while and found that the flickering lights floated back. He then tried to see if these were hot flames by slowly advancing his hand over the murky brackish water towards them holding a thin piece of paper. As he got closer, the dead hand of a zombie did not suddenly emerge out of the black mud, grab his wrist, and pull him into the marsh to a gruesome ghoulish death. Had it been me, the idea that this might happen would have dominated my thoughts, and any small noises in the dark forest would have amplified my growing terror. But Major Blesson was made of sterner stuff, and this allowed him to make an important discovery: he was able to light the paper in the flames. He concluded that the will-o'-the-wisps were caused by flammable gas bubbling up from the bottom of the marsh.

If marsh gas was the origin of these will-o'-the-wisps, a question remained. How did they self-ignite? Flammable gases react with oxygen in the air, but they need an initial spark to do that, and in damp wet marshes where would such a spark come from? Only an understanding of the chemistry of this ignition mechanism could explain the phenomenon, but this would take another 170 years (it is now understood to be caused by mixtures of phosphine and methane). Meanwhile other forms of mysterious light in the woods carried on amazing and frightening lost travellers.

One day I was walking back with my wife from the pub on a moonless night through the small paths of Dorset. The landscape of rolling hills full of small fields separated by hedgerows rivals the beauty of anywhere in the world in its sheep-nibbled bounty. At night though the greens all become grey and black and in the sunken lanes, which are covered by a canopy of hazel, maple and chestnut trees, you often cannot see anything at all, even your hand in front of your face. Some worries stirred inside me. It wasn't that we were going to get lost, or at least not seriously lost: we had walked this route home many times. But still the blackness amplified every doubt that popped into my head. It allowed some episodes from horror movies to creep into my mind. We had the idea of using our phones to illuminate the way. They provided a pool of light, but, somehow, seeing each other's startled and worried faces in the torchlight made things worse and only intensified our fear. Scenes from movies where innocents whose faces were lit by a torch were grabbed by unseen evil creatures came unbidden to my mind. So we turned off the torches and stood there, letting our eyes adjust and resisting the urge to grab each other in a protective huddle.

'*Whats that!*' said Ruby.

'*Nothing,*' I said.

But it was something.

A glint. A small sound: a branch cracking? The hairs went up on the back of my neck. Then we spotted a yellow-green light flickering in bushes not too far away. Tiny and pulsing, but definitely something. Just like in the horror movies, we felt unable to resist approaching it. The film audience at this point might have shouted a warning to us: 'Noooooo!' But as we came closer we realized that it was a small creature emitting this light, a glow-worm. They are actually beetles, not worms, that use light to attract their mates. They also use their light to attract prey, which they then devour. We had been attracted too by their mesmerizing glow but, thankfully, were too big for these glowing creatures to eat. Now we were up close we found a few more of these magical creatures shining like little stars fallen from the sky.

Glow-worms generate their light not through the production of heat, but by using bioluminescence, a so-called cold light. It is essentially a chemical reaction a bit like the reverse of photosynthesis. In photosynthesis plants use special proteins to absorb light and carbon dioxide and convert them into sugars while releasing oxygen. In bio-luminescence a chemical reaction does the opposite, using oxygen to create light and carbon dioxide. Many organisms have the ability to do this, such as some bacteria in the sea that create a luminous glow on the ocean. Crustacea, fish, squid and fireflies also emit this greenish light. These lights, once you know where they come from, are not creepy but rather fantastic and fairy-like. Our ancestors also knew about them and did not mistake them for ghosts, nor for the origin of the spark to light will-o'-the-wisps – which needed to be something hot.

Sit by a wood fire and you immediately realize that you are its slave, and you must feed it fuel to keep it alive and so to feed your fascination. As you look closely you see several elements to a flame. The core substance is a flammable gas that is released from the wood at high heat. As it emerges from the wood it meets the oxygen-rich atmosphere of air and emits a bluish light while it chemically reacts to produce hot invisible gases of carbon dioxide and water vapour. In a wood fire, with different currents of air rushing and swirling about, it is very difficult for this reaction to be complete, because all the carbon in the wood needs to chemically react with oxygen before it gets swept into the air. These small particles of unburned carbon become part of the air and glow red and yellow hot – this soot is the source of the fire's bright light. The particles are blown by air currents away from the hot wood until they are so far away that they cool down and no longer glow. This distance marks the edge of the flame. The reason the flame flickers, moves and sways is that the hot gas containing soot emerges at variable rates from the wood and is buffeted by air currents that are also dynamic. This description of the anatomy of a flame is not to downplay the magic of a wood fire, nor any other type of fire. But it does turn out to be very useful, because it allows a fire to be deconstructed into its parts, stored and transported as a gas, and piped throughout a city.

What is this flammable gas though, produced by hot wood? Charcoal makers have long known about it through their ancient practice. Camping out in forests they chop down trees and build large piles of wood which they cover with turf and earth. Down the middle of the pile is a channel to allow smoke to escape and an opening at the bottom to allow air into the pile. Lighting the pile creates a smouldering fire that turns the wood into almost pure carbon called charcoal. Just like the soot in the flame, this is due to partial burning with reduced oxygen. Because of this, most of the hot gas coming out of the wood cannot burn (react with oxygen) and so it emerges out of the top of the charcoal pile. The hot temperatures also liquify the resins inside the wood and these start to drip to the bottom of the pile. They too are flammable, but with the reduced oxygen inside the pile they cannot burn and so emerge as a viscous black river: tar.

Traditionally this black tar was extremely valuable. It is a mixture of long-chain carbon molecules similar in their chemistry to the tar extracted from crude oil. These carbon molecules are the remnants of the lignins and resins that hold a tree together when alive. They are extremely viscous and sticky and, being organic in nature, they repel water. Until modern times this tar was a vital ingredient for boat builders, who used this substance to caulk their wooden boats – its stickiness making it ideal to adhere and make a waterproof seal. It was also painted onto fabric to make waterproof sails and cloth – so-called oil cloth, the mainstay of weatherproofing until the modern era. Tar was also used in the building trade to waterproof houses and roofs. Its downsides were its toxicity and its tendency to degrade in extreme heat and cold (solved by the invention of vulcanized rubbers and plastics). The charcoal was sold as fuel for fires and to blacksmiths.

But what about the gas coming out of the top of the pile? Gradually, as chemistry established itself as a discipline during the seventeenth century, the invisible gases became the subject of serious study and were isolated and identified. It turned out that the explosive gas that killed miners in coal mines and the gas coming out of charcoal piles, as well as the gas produced in swamps, were the same gas. It was methane, the will-o'-the-wisp gas. In swamps bacterial decay of carbon-rich material such as wood and vegetation in the absence of oxygen created a carbon-based gas, with the formula CH_4. Coal deposits naturally produced this too, called 'natural gas', through a process that took millions of years deep in the earth. The charcoal-making process produced the same gas but much faster.

After its discovery many scientists and engineers wondered whether this flammable gas could actually be useful for something, for lighting perhaps, a function that had been carried out for thousands of years across the world almost exclusively using candles and oil lamps. Using a gas for lighting had some obvious advantages: being a gas it could be piped throughout a city and its houses, so the constant need to replace the oil and candles in lamps would be avoided. Perhaps, they thought, it would bring down the cost of lighting, which was a major expense in most households. Perhaps it might even be cheap enough to make street lighting affordable?

Street lighting was at the time, and continues to be, an important marker of the progress of a modern industrialized civilization. Towns and cities came first of course, but for thousands of years they were plunged into darkness as soon as the sun went down. This left the inhabitants of towns and cities stumbling in the dark unable to navigate their way home, especially if the moon was not in the sky. Accidentally plunging into a river, a cesspool or worse became a natural fear. The darkness was useful to many who wanted to hide their activities, which might be smuggling, thieving or mugging unwary travellers.

If gas street lighting could be developed, then no longer would the growing industrial cities of Europe be dark at night. No longer would a traveller lost in the web of dark streets of London feel a hand clamped over their mouth and be dragged into an alleyway to be robbed and have their throat cut. City streets with bright lighting would be safer streets and they would banish not just criminals and thugs but existential fears too. Well-lit streets and rational thought would go hand in hand, and a truly civilized city would be bathed in light. Gas lighting promised enlightenment in every meaning of the word. And yet to achieve it required taming the will-o'-the-wisp, the ghost of the woods, and bringing it into the service of civilization.

The first person to seriously try was a Frenchman and engineer called Philippe Lebon. A steam engine designer, he was accomplished at controlling and piping gases in these engines. He became interested not in the steam but in the other gases coming off the coal used in steam engine boilers. The chemistry of this coal gas was being better understood and as a keen inventor he saw the possibilities of using metal containers (retorts) to control the heating of coal and wood and capture the methane. He took out his first patent in 1799 and was so passionate about the idea that by 1801 he had the whole system working in the Hôtel de Seignelay in Paris. Five rooms, the gardens and the façade of the house were all lit by the flammable gas he produced and piped through the property. So marvellous was the spectacle of will-o'-the-wisps flickering away around every corner that the public happily paid three francs to enter and see the wonderland he had created.

Although impressed at first, the French were reluctant to embrace this future for lighting. Although the light was bright and they could see the advantages of not having to constantly replace candles or replenish the oil in lamps, there was a terrible stink. This smell was coming from the burning gas, which was odd because most of the gas was methane, a colourless, odourless gas even when burned. Hydrogen and carbon monoxide were also in the mix, but they too are colourless and odourless. It was small amounts of other gases that were causing most of the stink. The main culprit was hydrogen sulphide, which smells of farts and rotten eggs. This gas makes swamps smell swampy, but even in small quantities it is an extremely foul odour.

Was the will-o'-the-wisp warning those who attempted to civilize it about the consequences? The French didn't want to take the risk.

In Britain there was a different attitude. Riding high on the success of the Industrial Revolution there was an atmosphere of excitement about modernity, and a belief that technology could conquer nature's wildness. Money was also flowing as rich merchants were keen to invest in, and profit from, this new future. Boulton & Watt was the pre-eminent firm of steam engine builders that took up the challenge of delivering Britain's cities from the darkness of the night and of providing street lighting from gas. It was already expert at making gas technologies for handling steam, and also knew how to pipe and collect gases in pressure vessels. It had had a talented engineer as well, William Murdock, who had been playing around with a gas lighting system and so it was not starting from scratch. Its other advantage was that by the early 1800s coal was widely used for heating and cooking in British cities, and was delivered from the mines daily to cities and towns.

Coal behaves very similarly to wood when heated in a reduced-oxygen environment, producing not charcoal but coke, which was then sold as a fuel for heating or use in iron- and steelmaking. The liquid tar coming out of the bottom of the distillation was also valuable. The British navy was constantly experiencing shortages of tar to waterproof its ships. The best pine tar came from Scandinavia or North America, but supplies were often uncertain and having coal tar was an attractive by-product of the new gas lighting industry. In the end this tar was used for many things, including medicines, soaps and liquid fuels.

There was another factor: Britain was becoming richer and more ambitious in wanting to create a more 'civilized' life for its citizens. The safety of the public, especially of women when moving around the city at night, was seen to be vital for a civilized city. For women to participate fully in the life of cities, they had to fight not only for their right to an independent life but also to inhabit spaces that were safe. The spectacle of brightly lit shops and markets at night was also known to increase trade as shopping became part of the night life and entertainment of cities. So there was money to be made, and if Britain could show its rival France how to make this awkward gas lighting technology work, then so much the better.

British engineers forged ahead but hit technical difficulties time and time again. The problem of the stink was solved by taking the coal gas coming out of the heating process and bubbling it first through water, which dissolved any urine-smelling ammonia. The gases were then cooled to remove oil residues, which liquified. The gas was then bubbled through lime water (calcium hydroxide solution), which reacts with the rotten-egg-smelling toxic hydrogen sulphide gas and removes it.

They then had to solve the problem of storage. The gas was manufactured during the day by heating up coal and was then washed, cleaned and stored for the evening, when it would be used instead of candles. This meant building pressurized-gas storage tanks, called gasometers. These tanks needed to be huge, the size of buildings, and they needed to grow in size as more gas was collected but then shrink during the night as it was used up. Pressure regulation was also a problem: too low and the lighting was dim and wispy, too high and it would lead to incomplete burning in the homes and factories, and this meant that deadly toxic fumes would pollute the air. Gas leaks were yet another problem: metal pipework conveyed the gas around factories and homes, and under streets; leaks were caused by small cracks – and as we're talking about a gas, the cracks only needed to be *very* small – and so as the networks grew bigger, finding these leaks became harder and harder. Failing to find leaks was not just a business problem, as the valuable gas leaked away, but also a major safety problem, causing explosions and deaths.

Given all these technical and safety difficulties it's worth asking why the British engineers persevered. Part of the reason must surely be that dark streets were even more frightening and threatening to the average citizen than the occasional explosion caused by the gas leaks. It was 1812 when gas lighting began to be profitable in Britain, and such was the desire for this new illumination that it had reached every town in Britain with a population over 10,000 by 1826 and was adopted by all the major cities of Europe and America. It is truly an astounding rate of expansion of flickering will-o'-the-wisps on the top of lampposts.

Citizens of towns loved this gas despite the dangers: it transformed their lives and made city centres safer, more exciting and modern. Electric light didn't get rid of gas because by the time it arrived in the late nineteenth century, gas was also providing a very convenient form of piped heating that freed the householder from the need to make

Gasometers were still a big part of the architecture of
cities, as in early-twentieth-century King's Cross, London, 1935

wood and coal fires for cooking. Unlike a fire it could be
turned on and off, and when the gas boiler arrived in 1868
it provided hot water and eventually gas central heating.

Fast-forward to the present and there are millions of
miles of gas pipes under the ground in pretty much all
major cities in the world. It remains today the cheapest
and most convenient form of heating homes. The methane
gas we use now is mined in tremendous quantities from
deep underground and is utterly part of our life support
system. Moving away from our reliance on this fossil fuel
is a big part of the strategy for the future to achieve net
zero targets, but it will be one of the great challenges of
the twenty-first century to remove this methane from our
lives, where it is now deeply culturally embedded.

It is also deeply structurally embedded in our homes, towns and cities. Many of the gas pipes are old and corroding. Others are just poorly maintained owing to the expense and difficulty of maintenance because they are buried under roads, and so inspection is difficult. Barely a week goes by that there isn't a major gas leak in some city in the world. Paris, London, Beijing, New York have all suffered major gas explosions in recent years, destroying buildings and killing people. Night-time is the worst. When there is a gas leak in a building, a large quantity of gas can collect in the basement and mix with air; without someone awake to notice, it becomes a potentially deadly combustible mix. If someone makes the mistake of switching on an electric light at that point, they will most likely detonate the mixture, which can collapse the entire building, killing those inside.

One of the reasons why gas leaks are so disastrous is that they are hard to detect. A major obstacle for the adoption of gas had been the stink produced by the hydrogen sulphide in early coal gas. Once this was solved, another problem emerged: by the twentieth century the washing of gas was now so effective that it became hard to spot a leak, since the gas was completely odourless and colourless. Ironically the answer was to add back in an artificial smell that would make gas leaks instantly detectable. The smell used today is produced by adding a chemical called methanethiol into the gas. It is an organo-sulphur compound that smells of garlic/cabbage. It's instantly recognizable to anyone who has left the gas on – its marshy smell reminding us of its mischievous will-o'-the-wisp nature.

Gaslight the film

In 1938 Patrick Hamilton wrote a play called *Gaslight*, in which a husband tries to convince his wife that she is insane. He does so through a campaign of manipulating things in the home, and then when she notices them tells her that she is delusional and going mad. The title of the play alludes to one aspect of the campaign whereby the husband keeps dimming the gaslights in their home, and when she notices, he claims the lights are not dimmer – it is all in her mind. The play became a film of the same name in 1940 and this is how the term 'gaslighting' entered the English language to describe efforts by someone to manipulate someone's perception of reality.

Another horrible misuse of the gas in our homes has

been its role in assisting suicides. The great poet and novelist Sylvia Plath killed herself this way. She sealed the kitchen door of her London house with tape and towels and then turned on her gas oven and put her head in it. She died of poisoning from the carbon monoxide contained in the coal gas supplied to her house.

Carbon monoxide is colourless, odourless, invisible and lethal. It is a deadly killer because even in small quantities it will knock you out and then over several hours kill you as you continue to breathe it in. Thankfully, modern gas piped into our homes does not contain carbon monoxide and so is less deadly than gas used to be: the methane is not poisonous in itself and so it must fill the entire room, displacing oxygen, before a person will feel drowsy and collapse. Sometimes modern gas cookers and boilers do go wrong and create carbon monoxide due to incomplete burning of the gas. Every year there seems to be a news story about a whole family dying in their beds due to a build-up of carbon monoxide in their home as a result of a badly maintained boiler. That this odourless lethal gas will invade their child's bedroom while they sleep is the modern nightmare that all parents fear. Rationally we know how to combat it, not with prayer or meditation but with a carbon monoxide monitor. Yet like methane its very invisibility allows it to continue to feed our irrational fears and beliefs. If you do experience carbon monoxide poisoning call an ambulance. The paramedics will bring a canister of an invisible elixir that will restore your health, and it's our next gas: oxygen.

4. Elixir

Scuba diving is perhaps the most dreamlike of all conscious experiences. When you enter the water you leave your cumbersome self behind and enter a weightless realm. Scuba diving snips the cords of buoyancy and you slip below the surface. Now you are in a different world where you can move in any direction with ease. Fish swim past you without concern. The mask, wetsuit, flippers, breathing apparatus and large metal tanks on your back that were heavy and bothersome on the surface are now an extension of you – weightless and intuitive. You have a new existence.

The only thing that seems to link you to the surface is your need for air. Every breath is delivered through a mouthpiece. As you breathe in, it automatically opens valves controlling the compressed-air tanks on your back accompanied by a sucking sound. Exhaling results in a line of silvery bubbles which make their own way back to the surface. The sound reminds you how dependent you are on this gas, or rather one component of the gas, the oxygen in it.

Living creatures were not always so dependent on oxygen. When life evolved on the young planet Earth there were only trace amounts of oxygen in the atmosphere. The land was rocky and barren like Mars. There was a high concentration of active volcanoes with hot lava spewing and pumping out sulphurous gases into the atmosphere. Earth's oceans were blue but empty – there were no animals or plants, or any life at all. No seaweed, no corals, no fish, no dolphins. If you had been scuba diving in the oceans of young planet Earth, you would have been able to explore the geological marvels of the crystals and stalactite formations in underwater caves. You might have dived down to explore hot undersea volcanic vents and marvel at the spectacle of strange-coloured rocks and crystals growing there. Poking around at the bottom of the sea you might have found something extraordinary – a forest of tall red and turquoise chimneys rising up from the sea floor near hydrothermal vents. Over millions of years these self-building structures would evolve a self-organizing carbon-based chemistry sustained by volcanic energy. This was the emergence of life on the planet. It

yielded bacteria-like organisms that were to be the origin-
ators of all life on planet Earth.

Or so the story goes. How life started is still one of the
biggest questions in science. What we know for certain is
that at some point bacteria did become established in the
oceans. They used carbon dioxide gas dissolved in the water
to build themselves – hence the term 'carbon-based life
forms', of which we are one. One type of bacteria found a
way to capture energy from the sun using a chemical process called photosynthesis. Photosynthesis freed them from
volcanic vents as their source of energy, allowing these
organisms to go everywhere there was carbon dioxide and
sunlight. In a huge blue-green bloom of life they took over
the oceans, and in the process released oxygen as a waste
product.

Oxygen is one of the most chemically reactive elements
in the universe. It will react with pretty much anything. All
the oxygen gas around during the formation of the Earth
reacted with the iron, silicon and aluminium to form the
iron oxide, silicon oxide and aluminium oxide rocks that
make up the Earth's crust, such as silicate, granite and
haematite. We know that cyanobacteria bloomed billions
of years ago because we can see the evidence in the geological strata. There are layers of rusty iron oxide, indicative
of a large increase in oxygen which reacted with the iron
in the oceans to form them. This rapid increase of oxygen
is referred to as the Great Oxygen Event. This was 2.4 billion years ago. There was so much oxygen that even after
reacting with all the iron it could find, there was pure
oxygen left over. And then still more was produced by the

cyanobacteria, which meant that the oxygen bubbled up from the oceans into the atmosphere. This potent gas has been a major constituent of the atmosphere ever since.

Earth-like planets with oxygen atmospheres are very unusual in the universe. This is because, being so reactive, oxygen quickly gets used up in forming rocks or other chemical compounds, such as water. So having an oxygen-rich atmosphere, which is an indicator that the planet is constantly producing it, is a strong sign that there could be life on that planet. Any aliens examining our planet from afar would come to this conclusion too. Once it is in the atmosphere it supercharges evolution of primitive bacteria into more complex life forms. This is because breathing oxygen gives organisms access to the enormous chemical energy of this gas. More energy allows more life functions, eventually leading to the development of multicellular organisms with eyes, mouths and brains. Bigger oxygen-breathing organisms consumed others to grow even bigger, all the while breathing out their own waste gas, carbon dioxide.

The Great Oxygen Event was a pivotal moment in our history. After this there were two main variants of life on the planet, organisms that breathed carbon dioxide producing oxygen and those that breathed oxygen and produced carbon dioxide. This symbiosis is common in nature: it is a hallmark of a stable ecosystem, where one set of organisms depends on another. We rely on bacteria, algae and plants to provide oxygen for us to live, and they rely on us and other animals to create carbon dioxide.

We know from the fossil record that by 350 million years ago, in the Carboniferous period, atmospheric oxygen peaked at 35 per cent and life had now emerged from the oceans to flourish on land. There were vast forest swamps with giant ferns breathing carbon dioxide from the atmosphere and emitting oxygen. Flying in the oxygen-rich atmosphere had become a thing. There were huge insects such as dragonfly-like creatures with wingspans of half a metre. These giant insects breathed oxygen through their skin. Other big creatures on the land evolved sophisticated lungs for breathing oxygen, such as reptiles, crocodiles and later the dinosaurs that dominated the land. As oxygen levels decreased from 35 per cent to the present level of 21 per cent over the next few hundred million years, the giant insects were replaced by a huge variety of birds that

An illustration of life on Earth in the
Carboniferous period, 350 million years ago

evolved from the dinosaurs. Then, of course, came mammals and eventually we humans, who also inherited this dependence on breathing oxygen gas.

And so it is that oxygen is our elixir. We cannot go for more than a few minutes without it before shutting down and dying. It is our life force, a hunger that is billions of years old. We breathe without thinking. At night it fuels our dreams as our chest rhythmically rises and falls automatically. The gas is our life support system. But we can only currently breathe unaided on the surface of this planet. To address this limitation we have developed technology to allow us to travel outside the Earth's atmosphere and to explore other planets. This journey started a long time ago, with a desire to return to where we evolved, to the bottom of the ocean.

Breathing using our lungs under water doesn't work, even though water contains oxygen, because the maximum concentration of oxygen in water is only 3 per cent. But a bigger problem is that water fills our lungs. It is a liquid 100 times denser than air, which our diaphragm muscles do not have the strength to expel. As a result, the water gets trapped. Coughing is our instinctual reaction, but it is no good under water since we immediately suck in more water. Within minutes we lose consciousness due to lack of oxygen and that soon leads to death. It is the desire to get around this problem that has led to the development of a whole set of underwater gas technologies, such as aqualung diving equipment. These have allowed humans not only to explore our ancestral home, where the story of our relationship with oxygen started, but also yielded huge benefits for healthcare.

It has been symbolically important as well, creating a different relationship with oxygen, a more ritualistic one.

The first attempts to develop underwater breathing technology involved travelling to the bottom of the ocean in a barrel full of air. These diving bells are recorded as being used by the Greeks in the fourth century BCE. The problem, apart from being stuck in a leaky barrel that inevitably and unnervingly filled with water, was that the diver could not move around on the seabed. Nevertheless, it was remarkably daring. There they were, brave submariners, breathing at the bottom of the ocean watching the sea life through small windows as it swam by. By the eighteenth century this technology evolved into diving suits with lead shoes that allowed the diver to walk on the seabed. The air supply was pumped from the surface through pipes. A rope attached to the diving suit allowed the diver to communicate by tugging on it. It also provided the means to pull the diver back up to the surface. These umbilical cords to the surface were not ideal, since any interruption in the delivery of air to the diver caused by a leak would lead to drowning. Also, being on a tether meant that the diver's mobility was limited. The next step in the technology was to give the diver a container of air allowing them to be autonomous under the ocean. This required a method to pressurize the gas into a small tank and deliver it to the lungs. By the mid twentieth century an apparatus called Self-Contained Underwater Breathing Apparatus (SCUBA) was developed. It comprises a regulator valve attached to a tank of compressed air and a rubber mouthpiece. The action of breathing in opens the valve and

Early diving suit

delivers air to the diver at the same pressure as the sur-rounding water. Breathing out then shuts off the valve and the exhaled breath is released into the water.

One of the first technical hurdles to overcome for the scuba technology was the development of the air tank itself. The pressures needed are high because to hold enough air to allow an underwater dive of sixty minutes requires about 2,000 litres of air. The task, then, is to compress this air, which is the size of the interior of a small car, into the space of a cylinder which is about 12 litres in size. This creates an enormous pressure inside that cylinder. When you make such a pressurized vessel you are potentially making a bomb. If a small crack appears, then the internal pressure will rip the whole structure apart propelling pieces of sharp metal into anything and everything nearby. To make sure this never happens, thick high-strength steel is used. This makes scuba cylinders safe, but also makes the tanks heavy (15 kilograms for a 12-litre tank). This would be a problem if it were not for the marvellous fact of the buoyancy of water, which supports you and makes you effectively weightless. Thus underwater you do not feel the strain of the weight of the heavy steel cylinder on your back or, indeed, the weight of your own body.

With scuba equipment and a tank of fresh compressed air on your back you can jump into the sea and return to being a sea creature for a while. It is a wonderful feeling. You go back to the time before organisms like animals climbed out of the oceans and started colonizing the rocks and breathing air. You can dive among the cornucopia of fish and sea creatures, or dive down into the depths and glimpse the ethereal majesty of deep blue ocean.

There is one problem though: your lungs have evolved on the surface of the planet to absorb oxygen from air at atmospheric pressure. As you go down, the water pressure increases and so the pressure of the air (which contains 21 per cent oxygen, 78 per cent nitrogen, 1 per cent other gases) delivered from the scuba equipment also increases. The result is that you end up absorbing more of the other gases too, mostly nitrogen. Unlike oxygen the nitrogen is not metabolized by the body, and so it builds up. This is fine for shallow diving up to a depth of thirty metres. But if you go deeper, then significant nitrogen is pushed into your body tissues, which as you return to the surface starts to expand, forming bubbles. This produces acute pain in joints like the shoulders and elbows, swollen glands, fatigue and other more serious effects. It is called decompression sickness, or 'the bends', because it causes those affected to bend over in excruciating pain. It is a serious problem for divers and gets worse the deeper they dive.

Running out of oxygen is another big problem for divers. Anyone who has choked on food knows what it is like to suddenly be without the life force of oxygen. You try to breathe in, but you can't get the gas into your lungs. You make strange rasping noises and others around you ask if you are ok, can they pat you on the back? You can't answer because you can't breathe and so can't talk. Panic starts to set in. Your body knows that you don't have long. Almost immediately a set of biochemical signalling mechanisms takes over from your normal bodily functions and screams through your nervous system. The blood entering the lungs to collect the oxygen comes out of the lungs empty-handed. If this continues for a minute or two, the cells in your major organs run out of energy. Your liver, kidneys, heart and brain all start to fail. You become confused and delusional. Irreversible brain damage begins to kick in. Your memories, your knowledge, your jokes, your lopsided grin, all are erased one by one as the oxygen starvation continues until there is nothing left of your personality. Should you find yourself in this position and the emergency services are called, thankfully ambulances carry oxygen compressed in steel tanks using the same technology as for scuba diving, as does the mask that is put over your mouth as the paramedics try to resuscitate you.

Choking to death can happen relatively quickly, but you

can also die slowly of suffocation. Both are very horrible ways to die. Historically it has been viral respiratory infections that have been bringers of slow suffocation, the most recent of which was the COVID-19 pandemic, which has killed more than six million people globally so far. For most people the coronavirus causing the infection is fought off by their immune system before it does much damage. But for approximately 1 per cent of those who are infected, the virus ends up infecting their lungs. There the immune response causes the production of fluids which fill the lungs and so prevent oxygen entering into the blood. This causes a drop in oxygen in the blood, weakening the operation of all the organs of the infected person. The person becomes easily out of breath and tired. At this point their immune system is also weakened due to lack of oxygen. To be able to fight off the infection it needs help: it needs the elixir, oxygen. If you are lucky enough to live in a country with modern hospitals, then this elixir is on tap. Oxygen therapy administered via a mask is what you will be given. Indeed, at the start of the pandemic it was the only therapy available, and if oxygen gas had not been available millions more would have died. In some countries where demand exceeded supply due to a large spike in COVID-19 infections, there were heart-breaking scenes of people dying outside hospitals trying to get oxygen that had run out.

Oxygen gas is manufactured in a very simple way. Air is cooled using cryogenic apparatus to a temperature of $-183°C$, which is the boiling point of oxygen, at which point it liquefies into a very beautiful transparent blue liquid. This liquid is stored and moved around in lorries in which it is delivered to hospitals to be stored as a liquid under pressure. You may have seen them on roads and motorways. They are huge steel lozenges, giant versions of scuba diving tanks, rounded for the same reasons – to reduce stress concentrations and keep them from exploding from the pressure. These regularly travel the length and breadth of every modern country 365 days a year to hospitals. The oxygen is connected to a network of pipes distributed throughout the hospital which allows doctors and nurses to administer it to patients as and when they need it. It is a life support system in all modern hospitals.

In countries where oxygen is scarce or unavailable, the outcomes for patients are much worse. The technology used in these settings where there is no infrastructure for making and storing large amounts is called pressure swing adsorption. These machines suck air from the atmosphere and separate the oxygen. To do this it is piped through a material called a zeolite.

These zeolites are amazing materials. Unlike a solid rock

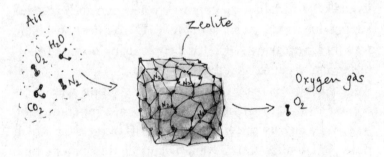

A diagram of a zeolite oxygen concentrator

a zeolite has a porous structure which makes it a rock-foam with lots of microscopic air pockets. Gases inside these air pockets can be selectively prevented from flowing inside the material due to the chemistry of the zeolite mineral, which forms sticky bonds with some gases, trapping them. In other words zeolites behave like a molecular sieve. Zeolites made from aluminosilicate occur naturally, and if you dig them out of the ground they naturally trap gases like water vapour. This can be released by just heating it up, at which point this seemingly dry stone erupts in clouds of steam. This apparent magic trick gave zeolites their name, from the Greek word 'to boil'.

Making artificial zeolites that are chemically tuned to adsorb different gases such as hydrogen (for hydrogen storage) is a trick materials scientists have been using for a while. In the case of zeolites for concentrating oxygen they work incredibly well. As long as a community hospital or home has an electricity supply, this allows a pump to suck in air and compress it. This compressed air is passed through the zeolites and the nitrogen molecules in the air are preferentially adsorbed. This turns the remaining air into almost 100 per cent oxygen, which is then removed from the zeolites and collected into a cylinder. The zeolites are then depressurized, which frees the nitrogen from the chemical trap of the zeolites into nitrogen gas. This nitrogen is then released into the air and the zeolites are now free to adsorb more. The cycle starts again with air being admitted into a cylinder and the zeolites start adsorbing once more. It is less efficient and more expensive than making oxygen through liquefaction, but it is a lifeline for patients in parts of the world where there are no bulk oxygen supplies.

For some people oxygen therapy isn't enough to turn the tide in the body's fight against the coronavirus. The lungs progressively fill with fluid. And even the enriched oxygen isn't enough to help the body function. Oxygen saturation of the blood dips further. The patient is unable to get enough oxygen and becomes too weak to breathe. At this point the doctors might try to put the patient onto a ventilator. This is a medical device which acts as a bellows, pumping air directly to the lungs. To do this the machine needs to bypass your larynx. This requires inserting a breathing tube into the lungs via the trachea. The patient can no longer eat, so a feeding tube is inserted. Nor can they talk, nor can they move as they are lying on their back covered in tubes and apparatus.

The procedure is terrifying and risky. The tube can damage the vocal cords and introduce new infections into the body. The pressurized air can also damage the lungs if not regulated properly. There is also a danger that too much oxygen will be delivered to the lungs causing a toxicity risk to the tissues.

Ventilators don't cure COVID-19: instead they buy the patient time. They keep the patient alive long enough for the body's own immune system to fight off the infection and for the lungs to be able to operate again. The fight is aided by other drugs that suppress the virus, but ultimately it is the person's own immune system that does most of the work. Is it like they are being plunged into the deep dark sea, a dream world where their subconscious is in charge. The masks and tubes are like scuba equipment, but the mind is drifting under sedation. Down, down they go into the interior of their subconscious, but this is more like a nightmare than a dream.

COVID-19 is not the only respiratory disease that ventilators are used to treat. Tuberculosis (TB) is a bacterial infection that has been spreading in human populations for at least 5,000 years but may be much older. It is known in the West as the disease 'consumption'. Like COVID-19, TB starts with a persistent cough and increasing breathlessness. Also like COVID-19 it affects other parts of the body too, such as joints, and causes headaches and body pain. TB sufferers experience weight loss and general wasting away, they become thin and also develop a pale complexion. It was thought that the disease was 'consuming' their body, hence the name 'consumption'.

For most of recorded history, it was viewed like cancer is today – as tragic but largely inevitable. Historically a third of fatalities were aged 15–34; half of those aged 20–24, giving consumption the names 'Robber of Youth' or 'Graveyard Cough'. Today many people can live with the infection, but the disease is fatal and has killed millions over the thousands of years it has been infecting human populations. Also, like COVID-19 the disease is transmitted through the air via tiny droplets as a result of coughing or indeed just breathing out. When a liquid is distributed in tiny droplets it is called an aerosol and it behaves like a gas, moving, flowing and hanging in the air for hours. Most importantly aerosols are invisible because the droplets are smaller than human eyesight can detect. If an ill person with TB or COVID-19 coughs or splutters you can appreciate they are ill and may infect you. But infectious aerosols of these diseases are emitted by the normal breathing of a stranger in a train, or a bus, or a shop, who may look perfectly healthy. The aerosols are colourless, odourless and transparent. You don't know you are being infected.

The rise of cities during industrialization provided perfect conditions for infection by aerosol. By the early 1800s TB was the cause of one in four deaths in England due to the cramped conditions of the poor. By the early twentieth century TB had infected 80 per cent of the population of the USA and was killing one in seven – typically those with other underlying heath conditions and weakened immune systems.

As with COVID-19 the main way to protect populations is to ensure buildings have good ventilation and to quarantine those infected. In the past the rich were sent off to sanatoriums outside of cities with access to plenty of fresh air. At the Pneumatic Institution in 1800 Humphry Davy tried laughing gas as a cure for TB, an example of the type of therapies being trialled by such hospitals. None of the treatments was effective and it was not until antibiotics were discovered in the 1940s and found to be effective against TB that the disease was brought under control. There are typically ten million new cases of TB every year, which cause three million deaths. This is the equivalent of a COVID-19 pandemic every year in terms of deaths. These deaths are preventable in the sense that good treatments for TB do exist, but people in poorer countries do not have access to the treatments, hence the death tolls experienced by those countries. Those suffering with TB die in a similar way to those infected by coronavirus, coughing and with lungs filling up with fluids, unable to take in enough oxygen and hopelessly fighting to keep the body's organs going. Almost as if drowning, they slip under the surface of consciousness. They are returning to a time before they were born. They are returning to a prehistoric world before our animal ancestors emerged from the oceans to make a home on the rocky continents. If they are to be saved they will need medical treatments including oxygen gas administered through the scuba-like equipment now standard in modern hospitals.

Oxygen technology is a modern engineering miracle, but it is not as remarkable as our lungs, which have evolved to capture oxygen from the air. Every breath delivers the elixir to our lungs, which operate without interruption for our whole lives. They don't even rest at night. Seemingly indefatigable they continue to deliver oxygen to our blood as we drift off to sleep, tipping us into our subconscious. Our rhythmic breathing keeps us alive as we submerge deeper and deeper into our dreams. There we experience a strange and surreal existence so different from our waking life. It has a quality perhaps like the underwater kingdom where we first evolved our thirst for this most remarkable of gases.

5. Musical

I can't play any musical instruments but this doesn't stop me being able to conduct orchestras. When questioned on this ability I shrug off any attempt to explain it. Only once in my life did I let doubt creep in. I was conducting a brass band as it made its way down the high street of a small English village. The musicians were ignoring me because I was only six years old. I was part of the crowd, and most of the musicians had their eyes fixed on their actual conductor. But one moustachioed tuba player locked his eyes on me and, while blowing his notes into the magnificent brass instrument, scowled at me. I dropped my arms immediately, along with my improvised baton. Somehow the tuba player managed to express that he was only kidding with just a twinkle of his eye, while also puffing his cheeks out to play the music. Delighted, I ran alongside him, conducting wildly and doing the occasional pirouette.

Since then brass bands have been my favourite type of band. They emerged from the coal industry, playing a type of music that celebrates these mining communities and the power they were providing for the Steam Age. Brass bands require no amplification, just human breath, and because the instruments are all portable the band can play and march at the same time. The brass instruments themselves, such as cornets, horns and tubas, are golden in colour, are bright and shine like the sun. Although they look and sound magnificent, they are also ingenious pieces of engineering that produce notes through a set of pipes connected by valves. This humble air valve is one of the most under-appreciated pieces of technology on the planet.

It is this invention that turns the exhalation of musicians into vibration. In the tubes of air in a tuba, trumpet, horn and cornet, harmonic notes are created that can be jaunty or serious, funny, farty, sombre or profound. Brass band music is the anthem of pneumatics, an exuberant field of technology that brought us the bicycle, the car, inflatables and the automatic sliding door, as well as the foam rubber that makes a sofa such a comfortable way to watch TV.

Our exhalation is not just the means by which we expel carbon dioxide from our lungs, it is our means of verbal communication through the spoken word, our laughter, our lament, our shouts, whispers and our song. This mix of the cultural, social and physiological roles makes breath our most personal gas.

Making musical instruments powered by breath starts with early humans collecting the horns of dead animals. These are essentially hollow tubes that are small at one end and wide at the other. Make a hole at the small end and blow down it while vibrating your lips and you will make a loud tooting sound. Pretty much all ancient civilizations from Africa to China seem to have adopted the horn as part of social and religious ceremonies. Many still use them today, such as the Masai people from East Africa, who use a horn from the kudu antelope. The horn shape acts as a natural megaphone, which was used to alert

The Masai kudu horn, still used in social and religious ceremonies

everyone within this extended earshot to take notice. The sound of a horn called the communities together, especially when they were separated in a forest hunting the very animals that they obtained the horns from. Still today, if you choose to hunt, the blast of the horn ignites our animal instincts. Rousing the spirit of an army to fight also became one of the roles of horns. These battle horns were the soundtrack of battles throughout millennia.

The horns themselves are treated as sacred objects, each one with a spirit that is expressed through the playing of the sound that comes from it. This sound is created by a resonating column of air inside the tube of the horn. The length of the horn determines the pitch of the sound. A longer tube produces a lower note, while a short tube has an increased pitch and so produces a higher note. You can tune the frequency of a horn by cutting it, or by carving holes in it to change the effective length of the resonating tube of air. An easier way to manipulate the note is to make a trumpet, which is a type of horn where the diameter of the column of air inside is kept largely constant. By changing the length of the tube you can produce any note you want. Then by flaring the end of the tube you create the megaphone effect of a horn as well as fine-tuning the harmonics. Of course, to do all this you need a material that can be crafted into a long tube. Traditionally bronze was favoured: it is malleable and can be shaped through hammering and moulding to create instruments with a high level of decoration and polish.

By making bespoke trumpets ancient rulers could demonstrate their power and magnificence through the sheer

Carnyx, ancient war trumpets of the Gauls between 60 and 30 BCE

fabulousness of the musical instruments and their loud sound. They could also make them of different lengths to create trumpets with different notes which when played together gave a triumphal crescendo. Metals such as bronze are not only aesthetically pleasing, shining like gold when polished, but the metal itself has an acoustic brightness which means that the sound resonates (this is also why church bells are made from bronze). This is different from natural animal horn or wood, both of which strongly absorb sound and give a muted mellower sound that is not as reverberant as bronze. With the functional purpose of a trumpet being to proclaim the arrival of a queen or king, or to begin a religious ceremony, or to signal to thousands of soldiers that the battle should commence, the bold sound of bronze was an advantage. The Age of Metals also brought many other advances in technology, such as the production of rock-cutting tools like chisels, which enabled stone cities, temples and tombs – the Egyptian Pyramids, for instance – to be built. Metal tools permitted the building of fleets of sailing ships and these made possible travel to and trade with other lands. Metal weapons allowed rulers to be powerful and to defend their cities, as well as to attack others. Bronze trumpets signified this sophistication and power.

The actual shape of the tube makes little difference to harmonic range because fundamentally it is still the same length of air, whether it is made into a straight medieval European trumpet or a coiled Middle Eastern horn. The air behaves the same way and produces the same notes. Coiling these metal instruments was useful though because it made them easier to hold when marching. Over time, the making of coiled trumpets favoured the use of brass rather than bronze. Both are copper alloys, and they look similar, but due to its zinc content brass is more malleable and easier to form into tubes of complex shapes. Using brass tubing also opened up the next modification: the ability to add extra sections of tubing by having joints.

Adding an extra section changes the length of the trumpet, and this allowed musicians to play lower notes on the same trumpet because the frequency of the note is directly related to the length of the vibrating air inside (e.g. longer trumpets can play lower notes). This allowed composers more versatility and range when writing music for trumpets and horns, although it was annoying for musicians, who were now called brass players. This is because these extra sections of tubing had to be added to the instrument while an orchestra was playing, which was time-consuming and laborious.

One such annoyed musician was Heinrich Stölzel, a German and an inventor. In the early 1800s he was employed by the army to play brass instruments. He resented the inconvenience of having to change the brass tubing on his horn. He realized that if the extra tubing was permanently attached to the horn but separated from the main instrument by an air valve, then by opening the valve he could effectively add the tube of air at the press of a button.

From an engineering perspective water valves had been around for a long time in the form of taps. Valves that controlled steam were being used in steam engines, while other valves were being used to control methane in street light technology. So this technology was ready to make an impact on musical instruments. What Stölzel invented was a valve for a brass instrument that was easy to operate with a single finger and that would close itself once released using a spring assembly.

Stölzel was not alone in solving this problem: fellow inventor and musician Friedrich Blühmel also developed a similar valve and they applied for a joint patent in 1818. These valves radically changed the musical range of brass instruments and allowed them to play full sequences of notes and to carry melody as well, bringing a fuller brass sound to orchestras. During the nineteenth century trumpets, horns, cornets and tubas were all transformed by the

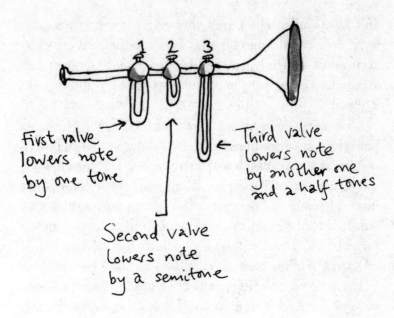

First valve lowers note by one tone

Third valve lowers note by another one and a half tones

Second valve lowers note by a semitone

How the Stölzel–Blühmel valve creates brass with a large musical range

Stölzel–Blühmel valve. When added to the trombone, which changes tubing length through a sliding mechanism, the brass band – a collection of instruments defined by a material, a type of tubing and air valves – was complete.

If Stölzel and Blühmel had known how important their valves would be to the development of brass bands they would have been delighted. They might have been pleased that the six-year-old me found the brass band music jolly enough to want to conduct. But even in their wildest dreams I don't think Stölzel and Blühmel would have believed that controlling air through valves would change personal transport. The first extraordinary invention it gave rise to was the bicycle. In fact, Stölzel and Blühmel were around at the birth of the bicycle in Germany – but as we will see, I am not sure they would have been impressed with the early prototype.

In 1816, two years before Stölzel and Blühmel submitted their joint patent application for brass instrument valves, Germany suffered a terrible famine. The famine was brought about by a temporary change in the climate which was dubbed 'the year without summer'. Germany was not the only country to suffer this – it was a worldwide event brought about by the volcanic eruption of Mount Tambora in Indonesia. The eruption was huge, and apart from the local devastation it caused, it also pumped large amounts of volcanic dust into the air. This dust stayed in the upper atmosphere for more than a year and created a hazy smog that encircled the Earth blocking out sunlight. The subsequent drop in temperature meant that some countries did not experience summer that year – hence the name. As you can imagine, there were a lot of doom-mongers who were not convinced by the rational explanation of a volcanic eruption. Church attendance in Europe and in many parts of America increased. The farmers in particular became more devoted to God, praying to reverse the change in climate.

But the prayers were not successful. On the east coast of North America, for instance, there were frosts even in July and August. Crops failed all over the world and there were major famines, including in Europe. German engineer and inventor Baron Karl von Drais witnessed this famine. He also witnessed that without grass or fodder to feed them,

The *Laufmaschine* (running machine), invented in 1817 by Karl Drais

thousands of the horses on family farms died. Many other horses were killed to feed the struggling families.

The death of the horses deprived communities of their dominant mode of transport. Karl Drais worried about this and wondered if he could invent a new form of personal transport. What he came up with made people laugh. He called it a running machine, and it looked like a cart that had been chopped in half. It had a wooden wheel at the front and one directly behind it at the back. In the middle was a seat where the rider could sit once they were tired of running. Running along a track on this machine was a bumpy experience and if you tried to use the momentum of the running machine to coast, on flat terrain it quickly came to a halt. The invention was no real help whatsoever to the farmers and merchants who, deprived of their horses, were having to walk everywhere and push their fruit and vegetables to market in carts.

But there was something intriguing about Drais's running machine. Anyone who had the patience to persist with it discovered something very odd and profound. If you started at the top of a hill and let gravity pull you down the hill, then not only was it fun, but for some reason the machine seemed to balance itself. You didn't need special circus skills to perform this feat of balance, anyone could learn to do it. This almost magical ability had something to do with speed because when the running machine came to rest at the bottom of the hill, it lost its superpower and toppled over.

No one knew why a two-wheeled device had special balancing qualities. In truth few people cared because going downhill was only ever going to be part of a journey, and sooner or later you would have to push the machine uphill and this was arduous. After the 'year with no summer' the climate gradually reverted to normal as the volcanic dust settled. Crops returned and real horses came back to being the main form of personal transport. But the strange running machine wasn't quite forgotten, and by the 1860s in France a group of blacksmiths and inventors added pedals, allowing a rider to self-propel along a road. A true mechanical rival to the horse as a means of personal transport had arrived.

It wasn't called a bicycle yet – it lacked a couple more inventions that would make it truly revolutionary. One of them was the drive chain, which allowed the back wheel to be rotated by turning the pedals. This made steering easier

Early bicycle designs, called velocipedes,
were pedalled directly using the front-wheel drive

and also gave better traction on roads. Gears were also introduced so that riding uphill became less of a struggle and
riding downhill could be even more exhilarating. But perhaps the most important innovation goes largely unnoticed
even now – that is, until you get a flat tyre on your commute
into work.

The wheels of the early bicycles were fabricated by carriage makers and so were made like carriage wheels of the
time: they were of wooden construction with an outer rim
of steel. This creates a strong and robust wheel but it produces a bumpy ride unless roads are completely smooth,
which they were generally not. The roads were mostly

rutted and full of stones and holes. The wheel design was also heavy and this made bicycles bulky and sluggish. Steel wheels were introduced with tiny thin steel spokes that made the wheel mostly air. The bicycle frame was also filled with air by using hollow steel tubes. But the problem of the uncomfortable ride remained. Famously the early bicycles were called 'boneshakers', and if you have ever ridden a bicycle with a flat tyre you'll have experienced this bum-smarting, head-aching, intolerable battering.

It was a Scottish inventor, Robert Thompson, who first had the idea to make tyres out of rubber filled with air. He developed them for horse-drawn carriages, but in 1888 another inventor, John Dunlop, thought they would work for bicycles. These pneumatic tyres reduced the weight of the wheel since most of it was filled with gas. They also made a bicycle more comfortable to ride because it was floating on a cushion of air. It couldn't work, of course: surely the thin layer of rubber in contact with the road would burst? While you do sometimes get a flat tyre, on the whole common-sense objections to the technology turned out to be wrong.

Air inside a tube of rubber when put under stress tries to squirt out, but finding no way to escape it compresses, which increases the pressure inside the tube. This pressure pushes back and it is this that holds the steel wheel, the frame, the pedals, the chain and indeed the 50–100 kilograms of rider on the bicycle off the ground. That a tiny thin tube of rubber can do this is nothing short of a miracle. When you sit on a bicycle with pneumatic tyres you are effectively riding on air. When you hit a bump or an imperfection in the road the rubber conforms to the shape of the bump and in doing so the elastic rubber absorbs energy; this stops the bump being transmitted up to the bicycle frame and to your bum and bones. One other vital component needs to be mentioned: the valve keeping in the air and allowing the tyre to be pumped up without it squirting out immediately. This valve, of course, is a descendant of the musical valve invented by Stölzel and Blühmel for brass musical instruments.

The pneumatic tyre turned the bicycle from a toy into a viable form of personal transport for the masses. It was no longer painfully juddering to ride but instead blissfully practical. It was also faster than solid steel because it was lighter, and by conforming to the shape of the road allowed for more nimble manoeuvring, especially around corners or in the wet. By 1889 cyclists using Dunlop's pneumatic tyres were winning all the now-popular bicycle races, and the bicycle was ready to fulfil its promise as a personal mode of transport.

The bicycle was also well suited to the new modes of urban life that were expanding for the populations of the world in the late nineteenth century. Riding a horse into a city for instance required you to have somewhere to put the horse: to shelter it, water it and feed it. The big cities were also filling up with horse poo and it was becoming a serious problem; for example, in 1894 citizens of London debated the coming 'Great Manure Crisis', which was a prediction that with the increasing use of horses the streets would become knee-deep in horse poo. In contrast a bicycle was nippy and comfortable and could be stored easily without fuss at your destination. It was also in tune with the age in another way. As the nineteenth century ended, steel and rubber were becoming higher quality and less expensive. The bicycle was thus an entirely modern invention that was more convenient and affordable. It liberated many who had no access to a horse and wanted to embrace modernity. The most liberated were women.

Up until the end of the nineteenth century most women were confined to the house or the farm. Most societies were patriarchal and insisted that a woman's place was in the home. In the industrialized world women did travel by steam trains and boats as well as horse-drawn coach, but they were chaperoned at all stages. When bicycles became popular it opened up a form of personal transport that was autonomous and challenged expectations for how women should behave. The bicycle itself became a symbol of rebellion against repression. Sitting astride a saddle was seen by many as being highly sexualized, and so cycling became a statement of women's liberation. The clothes women wore also changed in response to the need not to get the skirts of the time tangled in the chain or the wheels. Thus women wore trousers to cycle also as a symbol of emancipation and freedom.

The bicycle did not just become part of a political movement: for those living in rural areas the bicycle opened up a much greater range of experience for people to visit other villages and towns. This led to women being able to socialize more widely and attend the dances and parties being held further away. This had a measurable impact on the types of romances and marriages that took place. A study by the geographer P. J. Perry in 1969, using parish records in England, found that before 1887 77 per cent of marriages took place between people from the same parish. However, between 1907 and 1916, this had dropped to 41 per cent and marriages among people who lived between six and twelve miles apart doubled. In reviewing this the geneticist Professor Steve Jones concluded that the greater genetic diversity brought about by this change in distance between marriage partners was caused by the arrival of the bicycle.

The ease of riding of a bicycle is one of the reasons why it became so popular. Strangely its ability to self-balance has only recently been explained. At low speeds the rider actively needs to steer the front wheel to make sure the centre of mass remains in balance. But as the speed increases, the actions of different parts of the spinning wheels and the steering mechanism align to automatically balance the bike. This is not a gyroscopic effect, which is the force that keeps spinning tops stable. In the case of bicycles it only has a small effect. The self-balancing is not to do with the rider either. A number of experiments have been carried out to show that bicycles set in motion without a rider will self-balance even on bumpy roads. No, bicycles self-balance due to the geometry of the steering and position of the front axle. As a bicycle starts to tip over and fall, this creates a rotational force on the front wheel which turns it in the opposite direction to the fall, which steers the bicycle out of the fall and back upright. It might then immediately tip over the other way but as long as the bicycle is going fast enough, the wheel will now rotate to steer it upright again. This last point about adequate speed is important: stationary bikes don't self-balance because, although the wheel rotates as it falls, there is no forward motion to allow the wheel to steer the bicycle back upright. The self-balancing mechanism is simple but no less magical for that. To the rider it also feels like harnessing magic, and this seems to be one of the reasons why riding a bicycle is so wonderful.

Had Karl Drais been there with me and Stölzel and Blühmel that day when I was conducting the brass band, I think he would have been proud of the number of bicycles being ridden in the village and understood that they were recognizably versions of his original running machine. Stölzel would probably have patted him on the back and apologized for thinking it was a silly invention. The six-year-old me, jumping up and down with excitement at the presence of these inventors from the past, would have tugged at their jackets to get their attention and then I would have blown their minds by pointing to the cars whizzing past. They too have pneumatic tyres that are central to their operation. Although bigger, with more rubber, they are essentially the same as bicycle tyres. They even have the same air valve, which by the use of a small stick inserted into the valve, the six-year-old me had already worked out, would release all the air with a satisfying hiss.

The valve on almost all cars and many bicycles is called a Schrader valve. It was invented by an August Schrader and works by having an internal spring that continually pushes a rubber seal shut. This keeps the air in the tyre. A little metal stem pokes through the centre of the valve. When you insert a pump onto the valve to inflate your tyres, it presses down on the stem, which acts against the spring

A Schrader valve cap advertisement, *National Geographic*, April 1921

and opens the valve. To stop this happening accidentally and to deter six-year-olds, the Schrader valve comes with a cap, which was once celebrated and even deemed worthy of an advertising campaign in its own right.

In the twentieth century everyone went automobile crazy and garage mechanics had to unscrew a lot of valve caps as they mended the cars and pumped up the tyres. Soon they tired of this repetitive task and started to automate the pumping process using compressed air. This enabled them to do more than quickly inflate the tyres: it also created a new market for pneumatic tools. These tools could blast air and automate all sorts of mechanical tasks in a garage such as pressing tyres onto wheels. This compressed air also pushed tops onto bottles of beer, and Coca-Cola, and many other drinks. Factories became pneumatically powered and it wasn't just putting tops on drinks bottles, but pretty much any task that required mechanical assembly or fabrication.

This use of air as a mechanical tool is clever because the air itself is an almost limitless recyclable resource. A compressor sucks in air using an electric pump and compresses it into a small container, which increases its pressure. This container is connected through pipes to wherever it is needed. The system can be used not just as a power tool, but such is its versatility also to send messages. This is done by reversing the electric pump and using it to continually suck air out of a connected set of pneumatic tubes. Any message placed in a capsule and inserted into the pipes is then sucked along the system and can be retrieved at the other end. Such pneumatic postal systems were all the rage in the late nineteenth century in offices where paperwork could be transported from one office to another rapidly and with aplomb. In cities like London share prices written down on a bit of paper were whisked across the city to give maximum advantage to the receiving traders. Orders for fish and other perishable goods were sent this way too. At its height there were more than twenty miles of pneumatic tubes criss-crossing London carrying messages. All of it was made possible by an industry that fell in love with air as a material, a material that could be controlled by pumps and sophisticated valves. Next came the automatic sliding door. Through the wonder of pneumatics no longer would someone have to physically open a door. Instead compressed air would do the job, and in doing so make a satisfying hiss. This rendered door handles redundant and made the Tube trains sleek and modern. They first appeared on London Underground trains in 1919 and have been a characteristic part of the Underground ever since, not just mechanically but also acoustically. All those musical sighs and expletive bursts of sounds that accompany the experience of riding the Tube are courtesy of the pneumatic pumps and automatic valves on the trains. You can still hear them today on almost all trains, as they burble their message of salutation to travellers as they enter or exit.

The early twentieth century showcased another pneumatic technology that soon became part of everyone's life with its own acoustic signature – inflatables. The inflatable rubber ring was first invented as a lifesaving device for passengers on the steam liners – one gas technology providing a need for another as steam travel boomed. It was soon realized though that these inflatable rings possessed a cheeky playfulness that gave them another role in the leisure industry.

Throw an inflatable rubber ring into a lake and it floats in a way that seems to challenge you to try to sit on it. This turns out to be harder than you think. It generally requires multiple attempts, most of which result in you being dunked under the water and humiliated by what you thought was an inanimate object but now you think has a grudge against you. Any victory you win over the rubber ring is short-lived, especially if there is anyone else around who wants to get into the tube as well. Even very serious people find it impossible not to feel delighted by an inflatable ring. It is silly and irresistible in equal measure.

Of course, there is no reason to just stick with inflatable rings once you have realized their potential. No swimming pool is complete without an inflatable lilo to lounge about on, sip cocktails on, and occasionally slip off into the cool water when you feel too hot. Obviously I am not talking about my home country, Britain, here, as it is usually too cold to lounge in swimming pools, with or without inflatables.

In Britain the inflatable made its way indoors. First on the list was to re-create the feeling of pool lounging in the bedroom by replacing cumbersome and heavy sprung mattresses with this new modern technology. By the mid twentieth century inflatable beds became available, although this didn't go quite according to plan. The thing about an inflatable mattress is that the air pressure pushes equally over the skin of the inflatable mattress making it taut and almost solid until you sit down on it. This causes a pressure wave which quickly redistributes itself throughout the mattress. The upshot is that anyone else on that mattress is suddenly bounced off it. This anarchic behaviour could have been an exciting new feature of bedrooms, but the marketing teams never managed to persuade the public to ditch the traditional mattress – beds were for sleeping and sex and neither was enhanced by being unexpectedly launched into the air every time your partner made a move. This feature of course is the very reason inflatables are fun. The bouncy castle is the epitome of the genre, loved by children everywhere.

Designers scratched their heads about how to harness the exciting and bouncy quality of inflatables but at the same time make it less anarchic. It became clear that shoes could be redesigned to take advantage of compressed air. In the 1930s new plastics were created that had air incorporated inside them – these were foam rubbers. And it was Dunlop, the rubber tyre manufacturer, that brought them onto the market first. They are materials in which millions of air bubbles are effectively glued together to create a new material with all the springiness of a bouncy castle but in sheet form. Sports shoe manufacturers adopted them wholesale, giving a spring in the step to every athlete.

Trainers were transformed too: not only did the compressed air in the foam give a spring in the step, but they also reduced the shock waves travelling up to the knees and hips. As with the bicycle and car tyre, using air in shoes makes them lightweight and therefore the wearer is less fatigued lugging them around. Combining the cushioning and springiness of compressed air was the key to creating the most iconic trainer of all time, the Nike Air trainers.

Many other manufacturers of trainers have products with similar properties these days, but it is in the look where the Nike Air continues to win out – everyone can

Nike Air trainers – a pneumatic design classic

see you are literally floating on a cushion of air. This changes how you feel about yourself, how much energy you have, how cheeky you can be. Admittedly, this feeling is somewhat numbed today through our familiarity with them, but imagine how revolutionary they seemed when they were new on the scene.

I am not sure if Stölzel and Blühmel would have understood the importance of the trainers I wore that day as I conducted the brass band. Of all the modern things I would have shown them that were legacies of their development of the air valve, trainers might have seemed the most pointless to them. The bicycle, the car, the inflatables all require valves to operate. The air in trainers has no valve, the air is encapsulated, it cannot make a sound, indeed the very quietness of this type of footwear might have alarmed them (they are called 'sneakers' because of this). Stölzel and Blühmel were musicians and they invented the valve to create music. It is the music of brass bands that sings the praises of compressed air to anyone who listens. It is the anthem of a technology that gives us the pneumatic pleasures of bicycles, cars, trainers and, of course, the fun-loving inflatables that capture the delightful silliness of being a child.

6. Enchanting

My earliest memory of perfume is being kissed goodnight by my mum when she went on a night out with Dad. I breathed it in while lipstick smacked on my forehead followed by some vigorous wiping off by Mum and then she was gone. The perfume stayed for a while though, a sweet vapour wafting round my room. At this point I'd jump out of bed and run downstairs to witness their departure and might get a nuzzle from Dad and *slam bam*, the molecules of his powerful aftershave lotion would hit me in the face.

Was it Old Spice, the only perfume I knew that existed

for men? This was advertised heavily on TV and showed a manly man surfing a turquoise wave in a tropical location to the soft sound of triumphant music. 'You'll become yourself,' a deep voice stated. 'You'll find success,' they continued, the urgency of the music increasing as the advert interwove pictures of the waves, the surfing man and a woman with beautiful hair. Finally, 'The mark of a man!' boomed out from the TV, the music reached a climax, and visuals showed a surprisingly small white ceramic bottle with a ship on it.

This white ceramic bottle was not to be found in my dad's bathroom cabinet. I looked for it on one of these occasions when my parents went out for the evening. Instead there was an odd-shaped glass bottle that had a gold top and said 'Eau de Cologne' on the outside in dark lettering. This was more like it! I opened the top and splashed some on my face as I had seen in the advert but missed and got it in my eyes which stung badly. I spluttered too, enveloped in strong-smelling fumes of my dad. I was excited though and ran upstairs to jump on my brothers' beds shouting 'The mark of a man!' repeatedly and pretending to surf the waves until forcibly ejected from their rooms.

Gases in the form of scents and perfumes play a different role in our life support system from the gases such as steam and methane which provide power in the form of heat and electricity. Our sense of smell is our gas sense: we sniff the air to detect danger and to discover delights. It is an important part of our emotional life too, evoking long-forgotten memories, like my dad's eau de cologne. This spicy floral fragrance will bring him back to me instantly.

We live in a time when you don't have to be rich to wear such a scent, but this was not always so. Thousands of years ago, when the first perfumes were created in India, they were expensive and designed for powerful rulers. The expense was not so much to do with the rarity of finding fragrances such as orange or rose: pretty much wherever you are in the world you could find wafts of floral or fruity scents. The problem was how to bottle them.

Just opening an empty bottle near a rose and waiting for its scent to flow in doesn't work for two reasons. Firstly, the molecules responsible for smells are present in small proportions in the air, and so capturing air near a rose plant will only capture a tiny amount of the essence. Secondly, empty bottles are not really empty but full of air. When you try to waft the air containing the nice smell into an air-filled bottle, there is no room in there and it can't flow in. It is like trying to pour coffee into a cup full of water. For sure, some coffee will displace the water, but most of the coffee will just spill around the outside. The same happens if you try to get the rose scent into the empty bottle.

To capture a smell you need to find its source and concentrate it into a liquid. The hunt for these sources and their concentration into liquids are the first tasks of a parfumier. For instance, rose scent is contained in an oil secreted by the plant. This oil is volatile, which means that it has a low boiling point and so easily evaporates. It is this evaporation that creates the rose smell.

Many plants excrete smelly oils. They are called essential oils and there is a long history of extracting them using a technique called distillation. The flowers are crushed and mixed with water. This is then heated up in a beaker with a cooling tube attached which collects the steam and vapour containing the essential oil. Over thousands of years the process has been perfected to be able to capture a wide range of natural smells such as mint (menthol), cinnamon, orange, lavender.

You can try this at home. It is not hard once you find the basic equipment. A warning though: the oil you distil from the rinds of oranges when sniffed will not transport you to the lush orange orchard of your imagination. Essential oils don't smell quite as you expect them to. This is because substances are seldom smelled in isolation and it is mixtures that create the scents that we connect to memories and emotions, or for instance to sitting in a real orange orchard on a hot sunny day. Re-creating the complex mix of grass, earth, zest and warmth is the art of the perfume maker.

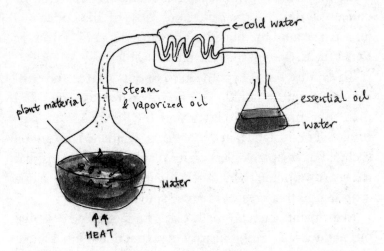

The basic extraction process is an ancient technique called distillation

These days the work of perfume makers is present everywhere in our lives, especially in our bathrooms. This is because we associate cleanliness with certain types of smell. It is not enough for something to look clean, it must smell clean too. We have all had the experience of stepping out of a shower and grabbing a fluffy white towel hanging on the rail in the bathroom. As the water happily drips down your body you use both hands to press the unblemished cotton towel to your face. But as you breathe in you smell stale mould. In disgust you quickly remove the towel from your face and examine it. It looks absolutely clean. You gingerly smell it again in case you were mistaken, but no, urgh, the towel is invisibly mouldy and you drop it onto the floor and look for another one. But there isn't one, this being your home not a hotel, which is also the reason for the mouldy smell.

And repeat, many times, in the history of washing.

To the rescue came soap, yes, but also the sun. Washing towels with soap gets rid of fungi such as moulds that grow in damp conditions. The detergent molecules attach to their membranes, break them up and make them soluble in water. This makes the towel clean but doesn't give it a smell. This is where scented soap powder and detergents come in. Essential oils and their modern synthetic equivalents are added to the washing detergents to give your towels and other washing that 'clean' smell. But what is that smell?

In 2020 scientists in Australia conducted experiments in which they took clean unscented towels and washed them with ultra-clean water and then hung up the towels to dry.

They dried them this way indoors, outdoors and also outdoors but not exposed to sunlight. They found that the action of UV light on wet cotton created small carbon molecules called aldehydes. You would almost certainly recognize the individual smells of some of these aldehydes: cardamom, citrus and rose. But together, in the proportions created in sunlight they have another smell, a combined smell: the clean-laundry smell.

The subtlety of the smell is hard to reproduce using additives to a washing powder, and in any case, what most people think of as a clean-laundry smell is the smell that they experienced as a child. This means that 'clean' is now defined as the smell of the laundry powder of their youth. Thus brand loyalty to washing powers and detergents is less associated with the ability to keep whites whiter, or colours brighter, but rather to their distinctive scent that emotionally connects them to home and family.

Despite the strong brand loyalty associated with smells there is a growing mistrust of additives in cleaning products. An increasing number of brands will now state if their product contains any of the known fragrances associated with allergies or eczema. There has been an increase in diagnoses of these conditions in developed countries but it is not clear why. One hypothesis is that children brought up in conditions of high levels of cleanliness might have immune systems that have a lower tolerance for certain substances. This has led to the development of detergents which brand themselves as 'fragrance-free' or unscented. These unscented brands will still clean your mouldy towels as well as the more popular brands, but they just won't leave a noticeable smell after the wash. Unless of course you dry your towels outside in the sun, when you won't be able to avoid creating the chemicals responsible for the clean-laundry smell.

Although we can identify the exact chemical molecules responsible for the natural smell of a towel drying in the sun, scientists still have no idea why we perceive the smells the way we do. In other words we cannot predict what it is about the chemical structures of these molecules that corresponds to their smell. This is not true of light. We know the wavelength of light determines whether we see blue, red or any other colour of the rainbow. Similarly, we know that the frequency of a sound determines the note we hear.

As regards smell, all we know is that there are some patterns in a chemical's structure that seem to be associated with particular odours. For instance, the family of molecules called esters smell fruity. You would recognize ethyl acetate as the fruity sweet smell of nail polish remover. In this family of ester molecules some are longer and some are shorter (depending on the number of carbons they have), but they remain chemically similar to each other and they still smell fruity. The type of fruity changes though: it goes from pear to melon, to apple and to banana. This seems to be solely due to the effect of the increased size of the molecule. By the time we get to an ester molecule with eight carbon atoms (octyl acetate) it smells of oranges.

To make sense of this effect of a molecule's size, scientists have been cataloguing the attributes of chemical

Carbons	Acetate Ester	Odour
1	$CH_3-C{\displaystyle {O \atop O}}CH_3$	glue/solvent
2	$CH_3-C{\displaystyle {O \atop O}}C-CH_3$	pears/solvent
3	$CH_3-C{\displaystyle {O \atop O}}C-C-CH_3$	melon/apples
4	$CH_3-C{\displaystyle {O \atop O}}C-C-C-CH_3$	apples/banana
8	$CH_3-C{\displaystyle {O \atop O}}C-C-C-C-C-C-CH_3$	orange

The family of ester molecules all smell fruity

molecules and comparing them with their smells. In one study hundreds of volunteers were asked to smell a range of substances and assess how similar they were to each other. This database was then correlated to weight, shape and how the electrons are distributed on a molecule. A complex pattern did then emerge between the smell of a molecule and structure, hinting that there might be some invisible logic to the similarities of certain smells. Artificial intelligence has been successfully used to correlate the database of chemical structure and smell, leading to claims that computers can smell. This is of course not true: smell is a human multisensory perception, not a calculation. The best AI can do is predict what humans will say a particular chemical smells like. It is very likely to get better and better at doing that.

A less logical but more popular approach to understanding smells is to do some cooking. For instance, put some sugar in a pan and heat it up. Soon the sugar molecules fall apart to form glucose and fructose, which then start to brown, which is a caramelization reaction – although you don't need to be told that because you can smell the gases that arise. The wonderful caramel smell is glucose and fructose turning into acetic acid (vinegar smell) and maltol (caramel), with a bit of furan (nuttiness). If you now add sliced apples, then the caramel smells are wafted up your nose accompanied by fruity ester molecules. Getting the ratio of sweet, sour and intoxicating maltol is the key to cooking the best apple tart in the world (in my opinion). You control the sweetness through control of the temperature, which determines the ratio of caramel. You control the sourness through the type of apples you choose, since different types of apple contain different amounts of citric acid, and it's the acid content which gives a dish its sharpness/sourness. When you smell you've got these ratios right, place a layer of puff pastry on top of the caramelized apples and put in a hot oven for fifteen minutes until the pastry has cooked and crisped up. Then remove it and turn the whole thing onto a plate. Now the caramelized apples will be face up to you with the crispy pastry underneath, and you will be hit with the full hot aroma of pie.

You must eat it hot, fresh from the pan, to get the full effect of this chemistry because temperature magnifies aroma. This is because the molecules in hot things have more energy and this allows more of the flavour molecules to become a gas, which creates the exquisite smell of this dish. But it has an even bigger effect inside your mouth, because of something called retronasal smell (through the nasal passage that links your mouth to your nose), which is the aroma of tart inside the mouth as it is being crunched. There are hundreds of compounds that are released by eating, and it is the smelling of these that delivers the sublime complexity of flavour when eating. This is why when you have a cold and your nasal passages are blocked, the pleasure of eating is vastly diminished.

Smelling requires the flavour molecules to interact directly with neurons from the brain that dangle down into the nose. They are protected by a layer of mucus. Inside this mucus there are special molecules whose job is to transport the captured molecules to the neurons. These neurons are connected directly to the area in your brain, the olfactory bulb, that detects smell.

Unlike the taste buds, there is no direct coding between olfactory system and the types of smell molecules. For instance, there are specialized receptors in your mouth that detect salt. They are looking out for salt and when they detect it, they send a signal saying they have found it. There are also specialized taste receptors that detect sweetness, bitterness, acidity and savoury. But, in the case of smell, there is no individual receptor for strawberries. Nor is there one for farts. The brain learns the difference between farts and fruit by paying attention to the pattern of receptors that 'light up' when interacting with different molecules. For instance, for the fruity ester molecules a similar pattern of receptors is active for them all, with small differences that change the fruitiness from pears to bananas. With only 400 types of receptor, the human olfactory system can distinguish between an almost infinite range of smells. This neural architecture is part of the reason why the smell of individual molecules has been difficult for science to decipher. But there is another reason, which is that once you have tasted something, your olfactory system remembers how much you liked it, or not, and creates pleasure associations with it. It is not an on/off thing but depends on mood and other associations. In other words, it is a multisensory experience associated with memory.

The first time I ate tarte Tatin was on an autumn evening in a small Victorian terraced house in East Oxford. I lived there while I studied for a PhD with my then girlfriend, Abigail. We were celebrating Bonfire Night with fireworks, and the damp night had the full bouquet of autumn leaves mixed with wood smoke and the occasional tang of gunpowder. My friend and fellow PhD student Andy Godfrey came over late with a bag of apples, a block of butter and a bag of flour. 'You've got sugar?' he asked me as he made his way through the party into our tiny, crowded kitchen and started to make the puff pastry by hand in a bowl, and using an empty bottle of wine to roll it out. Into the fridge it went, and Andy put a frying pan on our gas hob, nudging students drinking red wine out of the way. He heated up some butter and sugar until they started to caramelize, then peeled the apples into the pan and cooked them without stirring (this is the trick). Two minutes later out came the pastry from the fridge and it was flattened onto the frying pan with the apples underneath. The whole lot went into the hot oven for fifteen minutes, filling the kitchen with a fug of warm apple and buttery pastry. Andy leaned against the kitchen top drinking wine and opening the pot of cream which he fished out of his bag. Then the big moment: Andy took the pie out of the oven, flipped it upside down onto a plate, and the tarte Tatin was ready. Half an hour from start to finish and anyone who could find a plate in that crowded dark kitchen could have a bite of heaven.

All this memory is mixed up with my feelings about tarte Tatin, and so is also bound up in how it tastes to me. A shop-bought version doesn't taste good to me, nor does one from a restaurant. It has to be made fast and impromptu at home. Preferably it's not me making it, because, and here is another thing about smell, you can become so saturated with smells that you no longer notice them. This is called nose blindness, or olfactory fatigue. The reason for this is that your brain gets fed up with constantly receiving the same signal from the neurons and starts to ignore them to prevent overloading the nervous system. This is mostly a good thing, otherwise most of what you smell would be your own distinctive body odour. Nose blindness happens faster if you are cooking the dish because you inhale all the cooking smells, which is one of the reasons why cooks rarely enjoy their own food as much as their guests do.

Where you eat food also affects the smell and the flavour of a meal. I used to live in the high desert of southwest Albuquerque. The city is more than a mile above sea level. It is dry and sunny all year around. The dryness gives rise to the desert conditions but also means that it doesn't smell much. This is because rocks and sand don't have many volatile components, so not very many molecules are being ejected from their surface in the form of a gas. Since smell is our gas sense, and these rocks don't emit much gas, they don't have a smell, and nor does dry earth. But this all changes in a dramatic way when it rains. What water does, especially in the form of rain, is to disturb small pockets of air in soil and sand. This air is full of organic molecules that accumulate from the secretions of bacteria and fungi in the soil. When suddenly displaced by the rain, they rise up as gas and we get a sudden concentrated whiff of an earthy aroma that we recognize immediately as the smell of rain. There is a name for this smell: petrichor.

Petrichor smells depend on place. The petrichor of Albuquerque is a smell I can remember and even conjure up in my mind, it being so unusual in a place where the dryness enforced an absence of smell. A pungent sulphur smell, but not unpleasant like cabbage fart or swamp hydrogen sulphide, more gunpowdery as if the rock itself was being atomized by the droplets falling on it. It is so different to the petrichor of my hometown, London, where it rains almost every day like a sprinkler system on a timer. A muddy, grassy-green smell is released into the air every time it rains. That muddiness immediately brings to mind playing rugby on a Wednesday afternoon covered in mud.

The role of smell in labelling experiences, both pleasant and life-threatening, is well documented. The exact science of this is unclear, but in the mammalian hippocampus there are 'place cells' that are active when we go to a certain location. In the brain the signals from these cells integrate directly with smell detection, and this might be why some memories are evoked so immediately by a smell. My grandfather's flat smelled of a particular type of furniture wax, which whenever I encounter it takes me back forty years and I have to stop and sniff and smile. These days churches and temples are where you can go back in time to find that smell, along with incense, another ancient smelly material that comes from trees.

Some woods contain resins such as frankincense and myrrh which when burned emit wonderful smells. These are foundational incense materials that produce sweet-smelling smoke containing peppery and woody molecules: cresol, myrcene, and even limonene and vanillin. The association of incense with religious practice may have been due to its role in evoking emotion during ceremonies. In the Chinese tradition smells are an important way to communicate with the spirits of ancestors by cooking their favourite dishes, the aromas being the medium of communication. Smells linger in the air, mixing and uniting with the ghosts. Holy places that frequently use incense continually smell of it, as any movement disturbs the dust of incense and re-creates the odour.

My favourite incense is white sage. I discovered it in Albuquerque when my neighbours, who were from the Navajo Nation, invited me to take part in rituals, one of which was the sweat lodge. This involved three to four hours of sweating and singing in a small wooden construction, breathing the smoke of a fire and the smell of white sage burning. That they were so generous and open to invite me in to share their lives will always be associated for me with the smell of white sage.

My memories of Albuquerque come flooding back when I smell burning white sage. But there are other smells that I cannot access without actually being there, such as the smell of the house I rented up in the mountains among the desert cacti next door to my Navajo neighbours. That smell comes from a vast range of sources, including the petrichor emanating from the sand, the leathery wood smells of the timber frame, the musty and dusty furniture, and the synthetic pine smell of the fridge as I opened the door to get a cool drink. All gone for me. As is the smell of my parents' home and the home I grew up in. It would have been nice to capture that smell somehow, and there are many scientists in the world working on such a technique: a 'smell camera' that analyses the molecules in the air and records their composition and concentration. Such analysis is already possible and employed in many industries, not least the perfume industry. The difficult bit is then the ability to re-create the mixture of gas molecules with their associated fragrance on demand. It is a technology whose time will surely come.

The loss of the memory of particular smells is a normal part of life, but losing your sense of smell, and with it your ability to detect gases, vapours and aromas, is a far more profound loss. The sensitivity of our noses has evolved to be a protection system. It is one of the primary ways in which we avoid being poisoned. We smell food to see if it is off, to check if it is rotten and so likely to harm us. I sniff things from the fridge that have been there for a while. If I'm wondering whether the fish or meat is off, I'll give it a sniff. Similarly, that unmistakable smell of mould is easy to detect. We can detect the smell of poo with great sensitivity. The smell of urine puts us on alert, especially when we encounter it in a place where it should not be, such as outside our front door or emanating from a person on a bus. To meet someone who smells of poo or farts, however beautiful they look, is repugnant to most people, but not all. Even in this case the brain is able to associate a foul smell with pleasure, admiration or eroticism, and so make it appealing. The reality TV star Stephanie Matto took advantage of this in late 2021 when she started bottling her farts and selling them online. They became a big hit,

Stephanie Matto sold her farts online

and the demand was so great that she was able to earn $50,000 a week selling her guffs online.

The loss of our sense of smell is unusual and so it is often an indicator that something is wrong. In the case of the COVID-19 pandemic, loss of smell was a common symptom and a strong predictor of infection (some people experienced phantom smells while they were infected, such as 'burnt toast'). Smell disorders can also be an early predictor of certain cancers and even Alzheimer's. This loss of memory and the links to the emotional comfort that smells offer us, of home, of family, of the land with all its associated memories, is severed when our sense of smell is lost. This adds to a sense of dislocation from place and environment, which in turn leads to changes in personality. It shows the importance of our sense of smell for us to be rooted in community.

Smells are also important for building human relationships. Parental relationships start with smells, smells that are so fundamental, so part of who we are, that we don't notice them. Body odour is one of them. The odour is a combination of secretions from glands onto our skin via our sweat. These smells are affected by what we eat, by our health, by our age and by many other factors, such as the types of bacteria living on our skin. It is these bacteria that consume some of these secretions and give off pongs of their own. These smells can be pungent, such as the waft of body odour we associate with being sweaty. Once on our skin the molecules responsible for the smells evaporate from our warm skin continuously surrounding us in a cloud wherever we go. Many people try to remove this

cloud of smell by washing regularly or masking it with perfume and deodorant, but what if this cloud plays an
important role in the way we subconsciously communicate with each other? Could we be interfering in natural
ways of relating to each other by washing too much? This
is the strange world of pheromones.

Pheromones are substances that are intentionally emitted by organisms to control or influence another member
of the same species. Honeybee queens excrete pheromones into the air that entice other bees to mate with them.
Mammals such as horses, cats and dogs all use pheromone
odours to attract mates. They also mark territory with
pheromones secreted into their urine. Ants leave pheromone trails to allow others in their colony to find food. We
don't register most of these ant smells because they are
not intended for us and we don't have receptors for them,
but this invisible communication is going on everywhere
in nature. Insects, plants, animals are all sending and receiving signals in the air, much as we send our signals through
invisible microwaves that only our mobile phones can
detect.

But do humans use pheromones? Do we manipulate
each other through the excretion of smells which act at a
subconscious level? An experiment called the T-shirt test
has been the most popular way of testing the link. In one
study women wore T-shirts for three days, after which the
T-shirts were returned to the scientists. These T-shirts were
then given to heterosexual men to smell, who were subse-
quently tested for levels of the sex hormone testosterone.
The results showed that men who sniffed the T-shirts of

women who had worn them while ovulating showed higher levels of testosterone indicating that women may give off a pheromone that interacts with male hormones. However, these types of experimental result are contested, and conclusive evidence that humans use pheromones to control and influence each other has still not been found. This is telling because variants of the experiment have been repeated many times. Perhaps the reason for the lack of evidence is that modern humans have invented a more explicit way to attract and influence each other by smell – one that is controlled by our conscious selves rather than millions of years of evolution: the development of perfumes.

Perfume as a mixture or concoction is likely to have emerged from the practice of incense burning, since the etymology of the word is 'from smoke' in Latin. Methods for creating complex perfumes improved over time, especially when chemical processes for distilling ingredients came along in the nineteenth century. A theory of fragrance emerged which described a structure of perfume. 'Top' notes are highly volatile, which means that these are the gases that emerge easily from the liquid perfume and immediately make an impression, then fade within an hour. These are often citrus, pine and floral smells. 'Middle' notes emerge and overlap with them. They are often heavier, less volatile molecules – woody and fruity smells. These come out of the liquid perfume slowly and in lower volumes. In turn these give way to the 'base' notes, which linger and envelop a person for longer. They comprise smoky and animal smells such as musk and ambergris. The proportions of these, how they interact and extenuate each other and, perhaps most importantly, how they combine with the smell of our skin, give perfumes their character and an individual quality.

In other words, a perfume is like a menu with a starter, main course and dessert – a feast for the nose! For instance, the perfume Chanel No. 5 has top notes of bergamot flower and rosewood; middle notes of jasmine, rose and sandalwood, ylang-ylang and violet; and base notes of musk and ambergris, among some other significant ingredients, including aldehydes.

The popularity of musk smell in perfumes is at first

Advertisements for perfumes are modern spells cast over the population

surprising given that it is a creamy, oily secretion from near a gland next to a musk deer's genitals. It is a substance designed to attract other deer but seems to work on humans too. Ambergris is even odder. It is a black stinking bile from the rectum of sperm whales. Once it is expelled into the ocean it undergoes a transformation through the action of bacteria from the gut of the whale now cast adrift with the ambergris as their home and food. Their action and the effect of salt water and the sun produce amber naphtho-furan and ambrinol. These are heavy molecules that do not emanate into the air easily but, once there, linger and intoxi-cate with their sweet musky aroma. Many animalistic smells are irresistible for reasons that no one can explain. It's a kind of spell.

The power is not just that of creating a liquid which gives off a complex cocktail of gas molecules that interact with our olfactory senses: it is magnified a hundred times by advertising that associates perfume with success, beauty or some other attribute, such as 'manliness'. It is of course deeply naive to think that wearing a perfume will make you a different person: a person as beautiful or as rugged as someone in some advert. But it works. It is a billion-dollar industry that creates invisible but wonderful-smelling gases with addictive associations to images of a lifestyle or status we desire. It's a magic spell that works. Only we don't call it that any more, because we like to think we live rationally, so instead we call it 'marketing'.

Occasionally I smell my mum's perfume on the bus and can't stop myself turning around to check if it is her, even though I know it can't be. She has been dead for years.

I find it comforting nevertheless to be on the bus again with my mum. As a force of nature herself it suits her to be part of the air, capable of being a nurturing breeze or of blowing a gale. Such a breeze might be playfully ruffling the pages of this book now as you relax on a beach or under a tree in a field. Let these invisible air currents turn the page, it is their turn next.

7. Gods

When I was a kid, we watched the film *The Wizard of Oz* every year at Christmas. The ritual was instigated by my mum. The film spoke to her, she said. It terrified me. The winged monkey soldiers freaked me out most. Nevertheless, I watched it with her, hiding behind the sofa whenever the Wicked Witch of the West appeared.

The Wizard of Oz is set in the American state of Kansas, a very flat land. It is also a bounteous place where enormous fields of corn stretch into the distance as far as the eye can see. The flatness means that the winds blow steadily with

The *Wizard of Oz* twister about to pick up the farmhouse

little to obstruct or deflect them. This makes them dependable and constant except during the tornado season, when violent tornados create spiral funnels sucking up everything in their way, including houses, trees, fish, livestock and people. This is how *The Wizard of Oz* starts, with a twister picking up a farmhouse containing a small girl called Dorothy and her dog, Toto, and transporting them not just up into the air, but to the mythical land of Oz. Now as an adult when I watch the film, I see something else: a story about the forces of nature, and how we attribute god-like qualities to them as a way to come to terms with the power they wield over us.

We call hurricanes a force of nature, but our ancestors considered them a message from the gods – not a good message, an angry violent message, a message that destroyed crops, felled trees and pulverized towns. The list of wind gods from different cultures is large and includes Feng Po Po, the Chinese wind goddess who is depicted riding through the clouds on the back of a tiger; Aeolus, the ruler of the winds in Greek mythology, who could control the fate of ships at sea by creating storms; and, of course, the eponymous Huracan is the Mayan god of storms destroying everything in his wake.

Our ancestors found it important to appease these angry gods through ceremony and ritual. This was especially true of sailors. The small boats they built were vulnerable in storms, which would suddenly appear without warning and smash them to pieces. It is this unpredictable and seemingly irrational nature of winds that makes the idea of a wind god so convincing, especially in the face of a hurricane, when fifty-foot waves are crashing over a boat, and everyone on board is powerless to avert disaster: it is then that, even for the non-religious, prayer comes to mind.

I experienced this once when I was swept out to sea, aged eight, by offshore winds on a family holiday. My mum had packed me and my brother into an orange inflatable dinghy and then forgotten to keep an eye on us. By the time she looked up from her book we were a madly waving dot on the horizon.

Thankfully, after a while, the winds changed and we were blown further down the coast into another bay. The reprieve was a mystical moment for my brother and myself. We had been utterly helpless and yet we had been saved, and it was hard not to attribute this moment to us being special in some way. The thought popped into my head that a wind god had saved us and shipwrecked us into a strange land. Or, at least, it was strange to us because it was a nudist beach. Naked people pulled us shivering out of our inflatable dinghy onto the beach while we averted our eyes from their dangling bits. They were extremely friendly, and yet our British schooling had not prepared us on how to hug these wrinkly naked people who wore only sunglasses. In *The Wizard of Oz*, the heroine, Dorothy, is transported by the winds to a strange place and meets not nudists but Munchkins – tiny exotic creatures the size of children who wear bright clothes and have bald heads. Dorothy is disoriented and amazed in equal measure. But her overwhelming wish is to make it back home. It was our wish too, as we sat bewildered surrounded by naked people.

Not everyone regrets being transported by the wind to strange new places. For thousands of years sailing was the gateway to adventure for anyone wondering what was over the horizon. These brave people voluntarily took their boats to the water's edge, kissed their loved ones goodbye and set off over the horizon, often never to be seen again. It was an itch, an explorer's itch certainly, but the attraction was also to be in the embrace of the awesome and humbling forces created by movement of gas at a planetary scale.

The engine of the prevailing winds on Earth is the difference in temperature between the air at the equator, which gets heated to higher temperatures, and the air at the poles. Because of its lower density hot air expands, creating high-pressure regions. Cooler, low-pressure air moves to equalize the pressure and this air flow is what we call wind. North–south winds driven by temperature differences are an important driver of the wind systems on the planet but they are not the only one. There is another force that drives the planetary wind systems. It is called the Coriolis effect and it is to do with the fact that the Earth, along with its atmosphere of air, is rotating: we are spinning around our axis. The rotation means that at the equator the Earth is moving at a speed of 1,000 mph, while at the poles it has a speed of zero. The discrepancy creates a force called the Coriolis force. You will have felt Coriolis forces if you have ever been on a playground roundabout and tried to move from the edge to the centre: the rotational forces push you off course. The wind experiences these forces too at a planetary scale and they push it east–west.

The overall effect is the creation of planetary winds that are modified by the seasons (because this changes the pattern of heating and cooling from the sun) but remain pretty constant. They are called prevailing winds. They have a pattern in which easterly winds (trade winds) blow near the equator, creating the need for winds to replace the air displaced by blowing in the opposite direction at higher latitudes, and they are called the westerlies. The upshot is that if you are blown out to sea on an orange inflatable dinghy from an Atlantic beach in Europe, the trade winds will take you far out into the Atlantic Ocean and deposit you on the American coast. Exactly where you will end up depends on the sea currents but also on quite a bit of luck in not hitting a storm or being eaten by sharks.

If my brother and I had managed to be blown to the east coast of America in our dinghy, we would then have needed to find the right winds to take us back home. This would have required us sailing up the east coast of

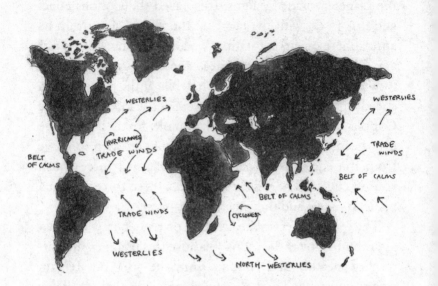

The prevailing winds are due to solar heating and rotation of the Earth

America to find the westerlies. With a bit of luck and a much better boat, we would have then sailed over the North Atlantic Ocean back to Britain, a distance of roughly 3,000 nautical miles taking 8–12 weeks to cross.

Thus the winds act as a kind of gas-driven conveyor belt which can take you round the world, and it is this insight that allowed early sailors to reach faraway places and return. But what if you want to trade up and down the east coast of Africa, for instance: how do you sail down one coastal route and return the same way? The prevailing winds would seem to make this impossible. However, sailors found an ingenious way around this problem which involves manipulating

the shape of the sail to take advantage of particular properties of moving air that are distinctly counterintuitive.

When the wind is behind you, then a sail will naturally
billow out and catch the wind, propelling the boat in that
direction. But if the wind is coming from either side of
your direction of travel you can still go forward by reefing
the sails. This shapes the sails into a curved surface which
is essentially an aerofoil similar to an aeroplane wing.
Winds blowing over an aerofoil create pressure differences
on either side. On an aeroplane this creates the lift force
that allows a plane to fly. On a boat the sail aerofoil creates
a force that propels the ship forward.

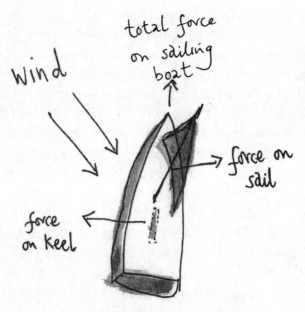

Sailing into the direction of the wind is
possible because of the aerofoil shape of the sails

Using sails in this way is all about balance. The force of the wind on the sails pulls the boat in one direction and the friction of the hull and keel in the water resists that pull. The resolution of these opposing forces takes you in the direction you want to go – if you are skilful enough. No amount of skill will allow you to sail directly into an oncoming wind, but you can sail close to the wind in a zig-zag fashion (called tacking).

Sailing like this is a marvellous feeling especially as the wind speed strengthens and the resolved force tilts the boat as it slices through the water at increasing speed. At the helm you are tilted too, and you must resist being pitched into the sea. As the wind tilts you even further over you slice through the sea with spray hitting your face. And if you are like me, you grin, sometimes with exhilaration and sometimes with fear. You are sailing the ocean by harnessing the elemental force of gas. It can take you to anywhere in the world – indeed, it was the wind that allowed us to explore the globe.

As some of our ancestors emerged out of Africa a million years ago, they did so on foot and then by boat, with the wind as their engine. Archaeological evidence shows that the islands of the Indian Ocean were settled by intrepid sailors 30,000 years ago. Over generations they navigated east, exploring Indonesia and the South China Sea, island by island. With sea levels lower than they are now, some of their voyages between islands were relatively small, between ten and a hundred miles, many of them within line of sight of each other. This is how the islands of Indonesia, Japan, Borneo, New Zealand and Australia were colonized 10,000 years ago. But then these explorers

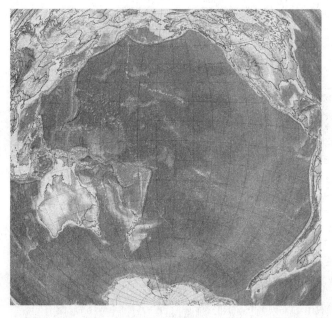

The Pacific Ocean

hit a vast body of water stretched out before them – the biggest ocean on the planet: the Pacific Ocean.

The Pacific Ocean covers a third of the planet's surface, it is 15,000 kilometres wide and there are no big land masses, only small volcanic islands separated by thousands of miles of blue ocean. Nevertheless, by 5000 BCE the sailors of Polynesia were using the wind to cross this vast ocean, eventually reaching Tonga, Samoa, Tahiti, Hawaii and even Easter Island. The story is one of skill and bravery not least because they had no way of knowing when they set out on their voyages that there was anything out there. They had no maps and no written language. If they ran out of food or water,

Traditional claw-sail ocean-going boat with outrigger. Boats of this design have been used by Polynesian navigators for millennia

or were shipwrecked by a storm, no one would rescue them. But still they sailed east, exploring and inhabiting what they found. They continued to do this for thousands of years, reaching Easter Island by 400 CE. Each expedition required sailing for weeks and months seeing nothing but ocean and an invariant horizon which was never closer and never further away. Was this just wanderlust? A vast number must have died but still the Polynesian sailors continued to explore. Why? Perhaps, it is speculated, they were trying to find the place where the sea ended, the place where the sun rose every morning, the home of the gods.

To accomplish their colonization of the Pacific Ocean,

the Polynesian sailors possessed extraordinary skills. On land you can tell how far you have travelled by looking for landmarks; you can also see paths and follow streams. Out at sea there are none of these, except when approaching land. For the vast majority of time in the Pacific Ocean, you see nothing. So the Polynesian navigators developed other ways of calculating where they were. One of these skills was a way of dead-reckoning their motion. It is akin to counting footsteps on land, where the counting allows you to know how far you have travelled. At sea there are no steps to count, and so it involves a calculation of the amount of time passed at a particular wind speed, taking into account ocean currents and wave patterns. At night they used the stars to determine their position, not by using astronomical maps or equipment but purely by eye and memory. Their memory was augmented by songs passed down from their ancestors. It is as if the wind itself could only be under-stood and communicated by the medium of breath.

In our family we had no oral tradition. It was not that my mum didn't try to instil one: she had a great singing voice, and frequently burst into song on long car journeys. We kids just thought it was embarrassing, especially when my mum sang the main tune from *The Wizard of Oz,* 'Over the Rainbow'. It's a song about finding a place where dreams come true, somewhere over the rainbow – a place where perhaps only the wind could take you. It is a very inspiring song, but it was lost on us, I'm afraid. We were little brutes back then, never content unless we were screaming and hitting each other.

Historically though it was this urge to fight and conquer

that pushed shipbuilding forward. Size is important for ships, because the bigger the ship, the larger the masts and sails it can support. Thus, generally, the bigger the ship the faster and more powerful it is. Bigger ships could also hold more sailors, soldiers and weapons, and so better defend themselves against enemies and pirates. Nevertheless, storms were still the biggest threat, and no ship constructed before modern times could reliably withstand the fury of a tempest at sea. That these storms were not random, but were instigated by angry gods, was the general belief of most seafarers around the world.

Seafarers prayed to these gods before going on a voyage. For particularly long or dangerous voyages the captains sometimes pledged to make a blood sacrifice or erect a temple to their gods at their destination. The remains of these built by the Phoenicians (a Bronze Age civilization from modern Lebanon) around 2000 BCE, and the other later seafaring nations such as the Greeks and Romans, are still scattered throughout the coast of the Mediterranean. The boats themselves were also protected with offerings and figureheads representing gods and goddesses at the prow of the vessel. In the Phoenician case the prows were carved to depict a horse-head with wings, giving the ships a life-like quality, so much so that records of the time refer to the 'death' of a ship when destroyed by a storm.

Before setting out on a voyage the captain or ruler, in the case of a fleet of ships, would consult a soothsayer to divine the will of the gods for safe passage and fair winds. This involved reading omens from a variety of signs, such as the flights of birds or the health of livestock in the spring. Even if the omens were good, things could go wrong, depending on who was travelling on a voyage. The Christian Bible recounts the tale of Jonah, who during a storm asks to be thrown into the sea to protect the others:

> And he said to them, 'Pick me up and throw me into the sea; then the sea will become calm for you. For I know that this great tempest is because of me.'
>
> Nevertheless the men rowed hard to return to land, but they could not, for the sea continued to grow more tempestuous against them. Therefore they cried out to the Lord and said, 'We pray, O Lord, please do not let us perish for this man's life, and do not charge us with innocent blood; for You, O Lord, have done as it pleased You.' So they picked up Jonah and threw him into the sea, and the sea ceased from its raging. Then the men feared the Lord exceedingly, and offered a sacrifice to the Lord and took vows.
>
> (Jonah 1:10–16)

Despite the belief that no boat, however strong, could provide protection from a wrathful god, the quest to make better boats continued. By the Middle Ages (1100–1400 CE) China had become a global leader in ship building, developing their so-called 'junk ship', which had a more rigid and more robust design with battened sails (sails made rigid by wooden reinforcement), integrated rudder and high-platform stern deck. The rigging of the Chinese junk was comprised of many smaller sails, each of which could be controlled individually and so optimized to take advantage of different wind directions and strengths. The mastery

A seagoing junk, identified as the treasure ship of Cheng Ho (Zheng He)

of the Chinese navy is illustrated by a famous set of voyages led by their general Zheng He.

In 1405 he set out on the first voyage with a flotilla of 317 ships carrying 27,000 troops and laden with highly prized Chinese goods such as silks and porcelain. The aim was to explore the oceans and assert military power so that Chinese merchant ships would be safe from pirates. The voyages were also concerned with advocating China's moral authority as the greatest civilization on Earth; a civilization blessed by the gods with peace and knowledge. China had maps of the world which revealed an important truth that despite the existence of trading routes on land such as the Silk Road, the oceans provided a much safer and faster method of transporting goods between the far-flung nations of the Earth. Safer, of course, because of China's increasing mastery of wind power.

It would take a few centuries for the Europeans to catch

up technologically with the Chinese. But by the fifteenth century there were a number of European maritime powers ready to take the sea route down the west coast of Africa around the Cape of Good Hope and up into the India and China seas. With the rigid class structure of European monarchies, there were very few ways for a person to get rich unless they were a merchant or a royal. But being an explorer was one of them. It was dangerous, and most explorers did not survive the trip, but those that did became rich and famous by bringing back spices and exotic stories of adventure. The appetite for these goods and stories meant that there was no shortage of explorers willing to take the risk, so as the years went by more and more ships sailed east.

Naval fleets became a powerful way to colonize and enslave other lands

There was a gradual realization by the European ruling class of the opportunity to project their power by ship across the whole globe through the colonization and enslavements of other lands. Guns, ships and aggression could win these monarchs wealth and empires much bigger than they could ever obtain by slogging it out on the battlefield in Europe against armies with similar capabilities. Much better and easier to go to lands inhabited by people without guns and colonize them.

Of course, colonization was not a new idea – it is as old as civilization itself. But between the sixteenth and nineteenth centuries the world witnessed a marked acceleration of sailing technology, which opened up the whole world to colonization by Europeans by sea. A pivotal moment was when the Italian navigator and explorer Christopher Columbus sailed west not east, and in 1492 landed in the Americas. He returned with tales of an exotic 'new world' inhabited by people unknown to Europe. He became a celebrity and soon the seafaring nations of Europe, principally the Portuguese, Spanish, French and British, all started to explore the Americas, and set up trading stations, forts, slavery ports and then colonies. It was a free-for-all that meant huge investment in ship building, navies, navigation technology and mapping. Back to Europe came incredible wealth in the form of tomatoes, potatoes, sugar, chocolate, tobacco, gold and silver – all delivered by wind power. Over to the Americas went enslaved people from Africa with the aim to use them and their descendants to work on plantations and make profits on luxury goods such as sugar and tobacco for the rich and powerful in

Europe – all made possible by wind power. Christianity too was exported by wind power to the Americas, with the aim of 'civilizing' and converting the indigenous people to worship a single almighty god.

By this time in Christian Europe, the sea and wind gods were heresy. But superstition continued at sea. Seafaring was too perilous to eradicate belief in such things. It was subject to the whims of the terrifying ocean storms or to a sudden leak in the ship's hull, either of which could send the ship to the bottom of the ocean. To be a sailor at sea was essentially to live under a suspended death sentence whereby your character and actions and those of your shipmates were thought to determine your fate. It was believed that birds carried the souls of dead sailors, and so killing one would bring bad luck on the ship. Similarly, saving someone from drowning was thought to be unlucky because it opposed the will of the sea gods.

Better maps and navigational technology were not deemed to anger the wind gods, which was lucky for the sailors because without them many more ships would have been lost. By the sixteenth century they were good enough to allow European ships to sail reliably to any destination, even theoretically around the world, yet until then no one had attempted it. The search was on to find a crossing through the American continent in order to pop out the other side, sail across the Pacific Ocean to China, India, Arabia and Africa, and then journey back to Europe. If no sea route across America could be found, some speculated they could sail north over modern-day Canada to reach the Pacific Ocean. Others thought the best bet was to sail south and find a route at the southern tip of Argentina, and then sail north to find the China Sea. But were any of these sailing routes viable? This was the question in a thousand taverns in a thousand ports the length and breadth of Europe by the beginning of the sixteenth century. Those who had sailed across to the Americas and back again had their opinions: they may have lost legs and arms and all their friends through disease, storms, fighting, drunkenness, drowning and starvation, but these seadogs still had a glint in their eye. The same questions were being asked in royal courts, although since the money and riches were rolling in from new colonies, there was little enthusiasm to send six ships and potentially lose them all just to prove a point that sailing around the world was possible. It took until 1519 before the navigator Ferdinand Magellan persuaded the Spanish king to invest in such an expedition in search of riches as well as glory. What seems to have swayed the king was the Portuguese dominance of the eastern sea routes to the lucrative Spice Islands near the Pacific Ocean. Spices such as cloves, nutmeg, pepper and cinnamon were the height of luxury in Europe and extremely valuable – some worth more than gold. Magellan promised that if they succeeded in going west, then a shorter and more lucrative Spanish-dominated route could be established.

A Spanish fleet of five ships left Spain on 20 September 1519. As is usual with these things, pretty much everyone died, including Magellan himself. The journey involved crossing the Atlantic to South America and being battered by storms in the South Atlantic Ocean. Then crossing the whole Pacific Ocean enduring starvation, scurvy, fighting and mutiny to reach the Spice Islands. In the homeward leg they kept sailing west, traversing the whole Indian Ocean to reach the Cape of Good Hope at the southern tip of Africa. From there they sailed north to reach Spain again. It was a trip using the prevailing trade winds, a journey of more than 40,000 nautical miles. They lost all the ships except one, and of 270 crew who set out only thirty returned. Sebastian Elcano, Magellan's second in command, got them back along with ten tonnes of cloves, an extremely valuable haul, which offset the costs of losing the other ships, although not the loss of life. Strangely they did not obtain glory, as somehow

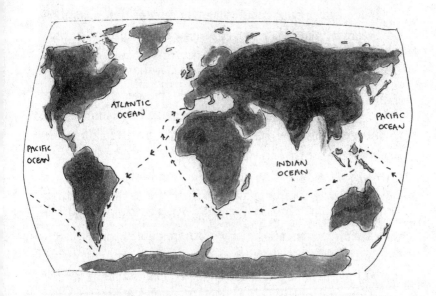

The route taken in the first circumnavigation of the Earth, 1519–22

Columbus is still the most famous navigator of those times, despite having achieved something less impressive.

Generations of sailors met the same fate, but gradually the belief that the winds were controlled by gods diminished as the science of weather forecasting became more accurate. Now satellites and computer models have replaced our reliance on soothsayers and astrologers for predicting the weather. We can determine with great accuracy the trajectory of hurricanes and typhoons, giving sailors and residents of cities several days' notice when they are going to be hit. This precision has its limits though: we can't predict accurately much further ahead than a week. This is because the invisible causes of the weather are not just complex – they are mathematically chaotic.

Chaos is a precise mathematical concept that describes how relatively simple systems can appear random and unpredictable. According to chaos theory this occurs when the behaviour of the system depends very sensitively on initial conditions. This means that small changes in, say, the heights of the waves at sea level can make huge differences to the evolution of a storm. It might be the difference between the storm blowing itself out in the early stages or growing to a huge hurricane. Such chaotic systems are fundamentally unpredictable because the cause and effect occur at many different scales, and being able to predict their behaviour means having accurate data at those scales. Gathering this information at the required accuracy turns out to be extremely challenging, and means that as we try to forecast further ahead, even very slight inaccuracies produce wildly different results. In practice this means that for the foreseeable future we will not be able to predict whether a hurricane or tornado will happen at a particular location next year, or next month. All weather forecasters can say is that they are likely to happen at certain times of year and in certain places, and once they do form, their trajectory can then be predicted reasonably accurately over the course of days.

This unpredictability of weather means that there is still a certain amount of mysticism in our attitudes to storms. We give them names and can't help but imbue them with a moral imperative. Mostly this is about punishment, whether it is the punishment of humans for causing global warming, or divine retribution. In 2005 many political and religious leaders declared that Hurricane Katrina, which killed 1,836 people and caused $125 billion in damage, was sent as divine retribution for the sins of New Orleans. The hurricane was classified as Category 5, the most severe category of tropical cyclone, with wind speeds of more than 170 mph. The speed of the winds is so high partly because of the location of the hurricane over water, which provides little resistance to absorb the energy. When hurricanes move onto land, the hills, trees and buildings provide that resistance and as a result the hurricanes become less intense. In the case of Hurricane Katrina, when it hit Louisiana it slowed to a Category 4 hurricane and in doing so ripped the buildings to pieces and laid waste to everything in its path, including cars, mobile homes and electricity infrastructure. Tragically many lives were lost, and it disproportionately affected the Black community, who lived in the worst-affected zones. New Orleans was not in its direct path, but it managed to devastate the city anyway by causing the levees that protected it to fail, which flooded the city, putting 80 per cent of it underwater.

Despite the devastation caused by winds, we can't survive without them. They are part of our life support system. This is because without winds there would be no weather: it is the winds that transport the moisture from the oceans to the land and deposit it there in the form of rain. Without this planetary engine that brings us our fresh water we could not survive on land: plants would die, crops would fail and desertification would ensue. Indeed, predicting rain to help farmers plan planting and harvesting is a vital part of our weather-forecasting service. It has mostly replaced the rain dances and other religious ceremonies performed by our ancestors during droughts.

When I think of the time I was swept out to sea with my brother, a part of me still thinks that something intervened to change winds to blow us back into the coast. I know it makes no sense – but can it just have been random luck? The answer is yes, but our irrational brains do not readily accept such explanations. Instead the experience of such events infiltrates our subconscious and our dreams, which direct our actions and beliefs in a more subtle way – as we see in the next chapter.

8. Dreams

The night before my first sports day I had a strange dream. I found myself in our garden. The flower beds and vegetable patch were gone and the grass was grey rather than green. Ahead of me was an enormous high jump – ridiculously tall – the bar as high as a building. And yet as I ran towards it and jumped, I sailed over the bar easily. Looking down, I saw my family below as small as ants looking up at me. I continued to float above them with complete ease. It was a good omen.

The next day at the sports ground I limbered up,

stretched, and looked down at my new trainers to check they were white and springy. They were. I was ready. Bouncing up and down on the spot I appraised the high jump. When the sports teacher gave me the signal, I bounded towards it with long strides and utterly failed to leap over it. My momentum carried me and the horizontal bar towards the blue landing mat, which I hit with a thud. I was astonished. It wasn't supposed to be this way, I complained, as I witnessed my fellow students leaping with ease over the high jump.

This was a good lesson, my mum told me in the car on the way home. You can't just dream yourself to be a good athlete, you have to train your body and mind. I was being a 'fat head', she said. But I had tuned out of my mum's usual lecture about how hard work was important in life. Instead I was trying to work out what my dream really meant. I had been to a birthday party on the weekend before my dream, there had been balloons there, some of which had escaped and floated up into the sky, never to be seen again. I quizzed my mum on the possibilities of balloons to help me float up into the sky. Why did they go up, I asked. 'That's the why,' she replied, which was her Irish way of ending conversations when I asked too many questions.

If you blow up a balloon using your breath, you are filling it with gas. That gas is very similar to the air you breathe, except it now has less oxygen, more carbon dioxide and water vapour, both of which your body has produced from the oxygen you breathed in. But such a balloon is denser than the surrounding air and can provide no help with flying. It will sink to the floor when you let go of it because density is what determines the buoyancy.

Balloons less dense than the surrounding air are buoyant and float upwards. The balloon density includes everything that makes up the mass of the balloon: the gas, the balloon skin and anything attached to it.

This explains why normal balloons sink to the floor: the air inside is the same density as the air around it, and rubber is denser than air, so overall the density is greater than air and the balloon sinks. The way to make a balloon float upwards is to put gas inside that is less dense than the surrounding air. Hot air is less dense than cold air and so hot air balloons can fly. But these quickly cool unless you have a method to keep them hot, and this requires a burner and fuel, which all increase the density of the balloon. It also means you will only stay up for as long as you have fuel. Another way to reduce the density of the gas is to use lighter atoms.

All atoms are made of sub-atomic particles. Most of the mass of an atom is in the centre, a place called the nucleus. It contains protons and neutrons. The oxygen atoms we breathe have eight protons and eight neutrons in their nucleus, which gives them a mass of approximately 16 atomic units. As a gas, oxygen binds together as a pair of atoms, denoted as O_2, and this gives oxygen an atomic mass of 32 atomic units, which is reasonably heavy for a gas. Carbon dioxide has two oxygens in addition to a carbon atom, denoted as CO_2, and is heavier still at 44 atomic units. Helium molecules are much less dense. Each atom has only two protons and two neutrons. Denoted as He, it has a tiny mass of 4 atomic units. Now we are in business! Helium balloons are much less dense than air, and so they float upwards.

After my disappointing sports day, and my mum's reluctance to engage on this topic, I asked my dad how many helium balloons I would need to attach to myself to fly. He told me about Lawrence Walters, an American truck driver, who had discovered the answer by attaching helium balloons to his lawn chair one by one until he took off. Forty-four balloons was the answer, and within a few minutes they floated him up to an altitude of 10,000 metres. Unfortunately for him the winds pushed him towards Long Beach Airport in California. As he floated towards the airport several commercial aircraft reported a strange flying object in their flight path. Walters had brought an air rifle with him with the intention of shooting the balloons to reduce his buoyancy and return to the ground. When he reached the height of the incoming aircraft he popped a few of the helium balloons, but then, unfortunately for him, the gun slipped from his grasp and tumbled to earth.

Lawrence Walters, a person whose mother, I suspect, also thought him to be a 'fat head', did descend eventually, crashing into power lines. He hung there tangled up but unharmed until the police came and arrested him for breaking the Federal Aviation Act. Afterwards Walters was unrepentant and, like me, cited his dreams as his major motivation: 'It was something I had to do. I had this dream for twenty years.'

Apart from the problem of controlling helium so that you can get up and down safely without an airgun, there is

Cluster ballooning inspired by the helium
balloon flight of Lawrence Walters in 1982

another issue with using helium to fly. The gas molecules
are so light they float right to the edge of the atmosphere
and then off into space. This is what happened to the
helium gas in Lawrence Walters' balloons once they burst. It
is also the journey taken by all the helium used in all the
children's parties that have ever taken place: it has gone into
the upper atmosphere. It has been lost. This is a problem
because there is not much helium on Earth.

The helium Lawrence Walters put in his balloons was
millions of years old – a by-product of radioactive decay
inside the Earth's core. An atom of e.g. radioactive uranium
has more than 235 protons and neutrons crammed inside its
nucleus, and this makes it heavy and unstable. So the atoms
split into smaller atoms, which are more stable. This is called
radioactive fission and it gives out energy too, which is how
nuclear power stations work. One of the by-products is

helium atoms containing two protons and two neutrons from the original uranium atoms. Once created in the core of the Earth, this helium bubbles through the hot magma and then through the microscopic gaps in the rocks of the Earth's crust. It is the end of a process that takes millions of years but eventually it gets quite near to the surface, collecting in places where fossil fuel gas reserves such as methane are also trapped. Thus the countries that mine methane are the ones which supply this by-product of the nuclear reactions inside the Earth, the helium gas we all use.

Although helium balloons are well known, the main industrial use of helium is as a refrigerant for vital hospital equipment such as MRI machines, and for scientific research. These require ultracold temperatures to operate: and helium's very low boiling point of $-269°C$ makes it ideal for this. As global demand for these technologies increases, so the demand for helium grows, often outstripping supply and raising prices. It is not clear when we will hit the limit of economic production, but what is clear is that the amount of helium we can extract sustainably per year is limited because the amount produced by the Earth is limited. So although using helium for balloons provides a shot of cheer for all concerned, it's a terrible waste – the helium ends up leaking out into the atmosphere and then into space, lost for ever.

But all is not lost for those of us dreaming of floating in the sky. There is an atom with even less mass than helium. It is the lightest atom in the universe. It has just one proton and one electron. It is hydrogen, a gas with an atomic weight of 1 atomic unit: it is 50 per cent lighter than helium.

An etching celebrating the first manned hydrogen balloon, 1783

Fill a balloon with hydrogen gas and it floats upwards with such urgency that the balloon yanks at your arm. Fill twenty of them and you will float upwards, but, of course, you need a plan for how to come down and also how to manoeuvre in the winds. There is another thing you need to worry about: hydrogen reacts with oxygen, often with a bang. Thankfully just mixing them together won't do it – you need an initial spark to produce an explosion. Sparks are rare in the sky, except of course when there is a storm brewing.

An appetite for risk is something that marks out all who take to the skies, but the pioneers of hydrogen ballooning, which began in the late eighteenth century, were particularly brave. True, the rewards in terms of fame and thrills were great, but those who did it often died. One of the most fearless was unquestionably Sophie Blanchard, who became the official aeronaut of festivals to Napoleon Bonaparte in 1804. Sophie was in the vanguard of daredevil balloonists. She did it for money and the adventure (crossing mountain ranges), but also because floating in the sky held aloft only by hydrogen was, she said, an 'incomparable sensation'.

Sophie used a chemical reaction to produce the hydrogen, which required mixing sulphuric acid and iron shavings. The hydrogen produced was directed into rubberized silk balloons through a series of tubes. Sulphuric acid in that quantity was fairly easy to obtain in the nineteenth century. Known as vitriol since ancient times, it had been used by alchemists for experiments but became more mainstream as a fabric dye and cleaning agent. Then it came to be used as a battery acid, and still is today (it is in the battery of every combustion engine car). It was dangerous because the hydrogen gas could ignite if anyone ignored the instructions not to smoke near the balloon. Vitriol is also corrosive to the skin, causing severe burns.

In the early nineteenth century, no fashionable party for the rich or high-profile national celebration was complete without a balloonist. But it was dangerous. Sophie lost consciousness on a number of occasions by going too high and suffering from a lack of oxygen, as well as almost freezing to death. After a while the public were thirsty to see more than just bravery and hear about feats of endurance across mountain ranges and oceans. So she started setting off fireworks from her balloon to amaze the crowd. They loved it. However, as is pretty obvious, this activity was extremely dangerous. A single misdirected firework would set the whole balloon alight in mid-air. Sophie was aware of this risk, and was warned many times about it, but she didn't seem to care.

On 6 July 1819 she ascended into the Parisian night sky in her hydrogen balloon wearing a white dress and a white hat with ostrich plumes, waving a white flag, and with a large number of fireworks onboard. As she ignited the fireworks, something went wrong. It is not clear what happened, but some reports mention that she went into a cloud in good shape but emerged moments later falling through the night sky.

Only once in real life have I found myself falling in mid-air hundreds of feet above the ground. Moments before I had been crouching in front of the open doorway of a biplane with the wind making a tremendous whistling noise. The same howling doorway that several other people had disappeared from moments before. None of them were pushed out of the door by our parachute instructor. He merely checked their equipment, attached their automatic pull cord to the aircraft frame, and then made the thumbs-up signal. Some of them jumped immediately and some paused for a moment before vanishing from sight. When it came to my turn I hesitated. This all should be fine, I kept saying to myself as I sat with my feet dangling out of the aeroplane. Thousands of parachute jumps are safely completed using this exact same equipment, I repeated to myself. I had seen the stats in the safety briefing. But still, every now and again someone did die: the news media

M. S. BLANCHARD celebre aeronauta
al momento del volo aerostatico da Lei eseguito in Milano
in presenza delle L.L. A.A. II. e R.R.
la sera del 15. Agosto 1811.

Sophie Blanchard flying in Milan in August 1811

reports these tragedies regularly. I looked down. There was an extra safety chute strapped to my front, I patted it. My instructor did the thumbs up again, this time in a slightly questioning way: was I too scared to jump? I knew that once I launched myself from the aircraft I would immediately free fall towards the ground, and while doing so I was supposed to count the numbers '1,001', '1,002' and

'1,003' out loud, then look up to see if my parachute had opened successfully. If it had (surely it would?), I could relax until I hit the ground 2,000 feet below. It was crunch time, a choice between a small but finite risk of death or never experiencing that 'incomparable feeling'. I jumped.

'ONE THOUSAAAAAAAAA!', I shouted, as I fell into the void. Then total silence. Just my heart pounding. I looked up belatedly and saw my parachute was open. I was floating with the wind, not against it, and so there was nothing for the wind to rub up against, hence no sound. No rushing wind. No howling. That was the most surprising thing. Being carried along with the wind meant I was not being buffeted by it either. I felt bird-like, poised, balanced and wonderfully serene. It was as close to my sports day dream as I had ever experienced, although I was not floating up but drifting down. Eventually the spell was broken when I saw the land below, which until this point had seemed far away, suddenly loom towards me. First I saw a patchwork of fields, then I could make out individual trees, fences and cows. It seemed like they were accelerating towards me. Then they were just fifty feet away and I knew I was going to hit them. But somehow I didn't – by luck I landed in the field between them, not breaking my legs or smashing my face in, but rolling on impact as I had been taught to do.

Sophie Blanchard did not land in a field, nor did she have a parachute. She didn't drift down but tumbled at a tremendous speed, striking first the roof of a Parisian house and bouncing off it, then slamming into the street below, breaking her neck.

This, sadly, was not unusual in the nineteenth-century

balloonist community. Many died attempting similar dare-devil feats, despite some having rudimentary parachutes. Countless more were blown thousands of miles off course by storms and ended up drowning in the oceans or falling to their deaths by crashing into mountains. Others like Sophie died because their balloon caught fire. Although the clear dangers put off most people, there were always more dream-ers to take their place. One such person had the idea of transforming hydrogen balloons from an adventurous pas-time to a means of mass transport: a German general named Ferdinand Adolf Heinrich August Graf von Zeppelin. His relationship with ballooning started when he was an official observer with the Unionist army in the American Civil War in 1863. There he saw the men of the Balloon Corps in action. They used hydrogen balloons tethered to the ground to observe the movement and position of the Confederate troops (at 500 feet of elevation they had a fifteen-mile view of the terrain). It was dangerous work because the Confed-erate army would spot them and fire directly at them. But it didn't put off Ferdinand von Zeppelin, especially once he himself got to ride such a balloon and felt the 'incompar-able sensation' of floating in the sky.

Zeppelin resigned from the German army in 1891 with the ambition to take the principle of ballooning and turn it into a steerable form of transport. He had in mind an air-ship using hydrogen to lift people off the ground and a motor with propellers to travel to any destination under its own power. This was vital to being able to steer and addressing the problem of being blown off course by the prevailing winds. The airships were to be like the newly

Hydrogen balloons in the American Civil War

invented submarines, except that instead of travelling through water, they travelled through air. The same physics that applied to controlling the buoyancy of submarines applied to airships, because both air and water are fluids (they have no fixed shape and they flow under pressure). In the case of submarines, to control buoyancy and be able to move up and down in the water requires adjusting the density of the submarine by pumping out water. For airships, buoyancy was achieved through adjusting the amount of hydrogen stored in vast volumes within the hull. For both forms of transport, an engine connected to propellors pushed them forward through their respective fluid environments.

There were a number of technical problems that Zeppelin had to solve to turn the prototype airships into viable commercial aircraft, most urgently the materials for the hull of the balloons. Early balloons were made from flexible fabric and ropes and were inflated into a round shape. They had no way to control their direction of travel and so moved with the wind. Steerable balloons were developed in the mid nineteenth century as cigar-shaped balloons with steam engines. They carried very few passengers, had limited power, and were easily blown off course. As the science improved, it became increasingly clear that airships needed to maintain their shape as they flew. The most efficient way to do this was to create a rigid hull for the balloon made from a jointed framework. The material for this lattice structure needed to be strong, stiff and low density. Wood fitted some of the design requirements, but it was not strong enough. Steel was considered but was much too dense. A new metal came to the rescue, the outrageously light metal, aluminium.

Unknown to industry and really a lab curiosity at the time, aluminium had an extremely low density, which made it very appealing. The problem for Ferdinand von Zeppelin was that pure aluminium is not strong. Could an alloy of aluminium be the answer? A German metallurgist called Alfred Wilm, unknown to Zeppelin, solved the problem in 1903 by going on holiday. He had been adding small amounts of different metals to aluminium to see if this made them stronger. He had good grounds to believe it was possible, because steel is made 400 per cent stronger by adding just 1 per cent carbon to iron. Carbon didn't work for aluminium, but what would the magic ingredient be? He tried many alloys and mechanically tested them to see if their strength had improved. He made no progress at all, and when small alloying additions of copper made no difference, he gave up and went on a boating holiday. On his return to the lab a week later, he saw the failed aluminium–copper alloys at the side of the lab, but instead of throwing them in the bin, he decided to test them one more time. Bizarrely he found that in his absence their strength had increased a hundredfold. They were now as strong per weight as steel.

He thought it must be a mistake and repeated the experiments many times, but kept finding the same strange effect. At first the alloys were weak and he could easily bend them with his hands. But then over a period of days they became stronger and stronger. He had discovered something we now call age hardening, a process by which a tiny set of crystals grow inside a metal, reinforcing it and so making it stronger. That metals are not just blobs of stuff but have complex internal architecture, akin to the cells in your body, was already known to metallurgists. What was also known was that you could harness mechanisms inside metals to grow different crystal structures to make metals harder or softer – this is how blacksmiths use a forge. By hitting steel with a hammer, the blacksmith not only shapes the material but changes the crystal structure, and by doing so increases the strength of the steel. But Wilm's aluminium was doing this on its own, without being hit at all. It was a key scientific breakthrough. This piece of magic made aluminium the metal for flight because now the dreamers such as Count von Zeppelin could optimize airship design for the airline he started in 1910.

Of course, there was always the issue of the explosive hydrogen gas used by these airships, which by this time were called Zeppelins. A number of the early versions crashed and exploded, but this didn't stop the engineers developing new and better versions. They found these cloud-faring machines too appealing and too marvellous to let go of. Like Sophie Blanchard and the balloonists that came before her, Zeppelin was in love with floating in the air, and he had a dream of a civilized future where humans travelled the globe in balloons. But as so often in life, one person's dream is a nightmare for someone else, in this case for the inhabitants of the cities of Europe who were about to experience a completely different future.

The First World War broke out in 1914, and the German army found itself, thanks to Count von Zeppelin, the only European nation with airships. With a top speed of 85 mph and the ability to carry two tonnes of bombs, they were a new weapon that could potentially change the outcome of the war. In particular, they provided a way to terrify civilian populations: night terrors became real terror.

On 6 August 1914 the Germans used the Zeppelins to drop bombs on the Belgian city of Liège. It was the first time in history that a city had been attacked from the air. Attacks on Paris followed, and by 1915 Zeppelins were taking off in Germany and navigating silently to hover over London under cover of darkness. The soldiers onboard looked down at the brightly lit busy streets below and dropped bombs on them. More raids followed, gripping Londoners with terror. Gunfire was not much use from below because bullets went straight through the canopy of the Zeppelin, causing hydrogen leaks but nothing serious enough to radically reduce the buoyancy of the airship. It was a similar story when British aircraft were deployed in

Zeppelins attacking the city of Calais in the First World War

response. No, what was needed was to be able to shoot fire that could ignite the hydrogen. In the end a special bullet containing nitroglycerine was developed. This didn't stop air raids though – the night-time attacks on civilians continued, by Zeppelins and newly developed aircraft. By the end of 1915 both sides were bombing each other with the aim of terrorizing the inhabitants of cities.

Once the war was over, civilian airships took to the skies and fulfilled Count von Zeppelin's dream. The airships got bigger, thanks to the age-hardening of aluminium, and their range was extended, enabling them now to cross the Atlantic. These enormous airships stunned the world. Perhaps this was partly because of their beauty as they hung

The *Graf Zeppelin* visiting America in 1928

in the air, defying gravity only a few hundred yards above people's heads. These slow-moving giants became a kind of new creature of the sky.

In 1929 a Zeppelin circumnavigated the globe for the first time, starting in New Jersey, in the USA, travelling to Germany, Tokyo, Los Angeles and then back to New Jersey. Routes to Brazil opened up next, and by the 1930s Zeppelins were carrying fee-paying passengers. Such was their popularity that when the Empire State Building was constructed in 1930, the idea that it would serve as a docking station for airships was promoted by its owners. It was a dream of city living in which inhabitants never came down to ground but inhabited the skies.

The biggest airship of them all was the Zeppelin called the *Hindenburg*. This airship was the size of a football stadium and a home in the sky for approximately fifty staff and fifty passengers. The passenger areas were luxurious, with lounges, writing rooms and dining rooms. Despite the large quantities of hydrogen gas onboard, it was felt unacceptable to ban passengers from smoking as they flew around the world. So they were also provided with a pressurized smoking lounge to ensure that hydrogen couldn't leak inside and ignite. For safety reasons the passengers were forced to extinguish their cigarettes and pipes when leaving the smoking cabin. The *Hindenburg* made seventeen return trips across the Atlantic in 1936. The journey time from Germany to the USA was fifty hours, which was by far the fastest and most glamorous way to travel. Travelling by train and steam ship, in comparison, took at

The *Hindenburg* disaster in 1937

least a week. It was expensive though, costing $400 one way (£23,000 in today's money) and therefore affordable only for the rich and famous.

On 6 May 1937 the airship era came to an end when the *Hindenburg* caught fire while docking in Lakehurst, New Jersey.

It had left Frankfurt on 3 May bound for the United States and had crossed the Atlantic en route to its docking station, when the captain was informed of thunderstorms. He delayed his approach until they had cleared but had some trouble keeping the airship in level trim for docking. Mooring lines were dropped for the ground crew to secure the airship, but a small flickering flame was spotted at the back of the ship. Then there was a detonation and the airship burst into flames. The moment was described on live radio by Herbert Morrison for the Chicago radio station WLS, which is still one of the most heartbreaking live descriptions of any disaster (the broadcast is available on the internet). He starts off describing the docking of the most magnificent airship in the world, when he spots the flames: 'It's burst into flames, and it's falling, it's crashing!' He continues to describe the airship turning into a furnace until he says, 'Listen, folks, I'm gonna have to stop for a minute because I've lost my voice. This is the worst thing I've ever witnessed.'

Thirteen passengers and thirty-five aircrew died. Amazingly, the rest of the ninety-seven people onboard survived, somehow protected in the cabin as it crashed to the ground and the hydrogen burned around them. There are many theories as to what caused the hydrogen to ignite, including lightning, a fuel leak and sabotage (this was a Nazi ship often used for propaganda). There is no consensus, although the official crash investigation concluded that a spark, some-how caused by the build-up of static charge, ignited leaking hydrogen, which mixed with oxygen in the air.

This was not the world's worst airship disaster in terms of number of deaths, but it was the most high profile, as the horror was captured on film and by broadcast journalists at the scene. The footage was communicated around the world, and the ensuing public nervousness about safety brought an end to the commercial viability of airships.

There are a number of memorials to the first era of air-ships, but perhaps the most enduring is the British rock band Led Zeppelin, who used the image of the *Hindenburg* disaster on the cover of their debut album. Apparently they chose the name to represent a kind of apocalyptic tri-umph. They associated themselves with the glorious ascendancy of hydrogen airships, brought down to earth like a lead ('led') balloon. In doing so they tapped into the feeling of post-modern-malaise Britain in the early 1970s, when it was gripped by economic depression and strikes that brought the country to a halt. It was a rebellious sound that resonated strongly with my brothers and me, feeling trapped and made miserable by the drudgery of school and the brutality of our teachers, who hit us with canes pretty much every day.

Looking back on my childhood now, I see my many floating/flying dreams as my rebellion against all of that. Against the oppression and cruelty of school, against the decay of Britain, and even against my mum's work ethic. I wanted to be allowed to dream, to dream myself to be a great high jumper, or anything else I could imagine: not to be trapped by the logic of hard work or even gravity.

Those flying dreams return occasionally in my adult life and still make an enormous impression on me, I find them nourishing, amazing and thrilling. I imagine that Sophie Blanchard and Count Ferdinand von Zeppelin had similar dreams.

Perhaps today's balloonists have them too, because ballooning did not die with the *Hindenburg* disaster – in the face of repeated failure and tragedy, people continue to float into the sky. Hot-air ballooning is thriving and every year thousands of balloonists ascend into the clouds. High-altitude balloons also play a part in weather-forecasting systems, being released many times a day to help predict storms. There are as well several commercial balloon flights that will take paying passengers (safely using helium gas) to the edge of space. They advertise themselves as low-carbon alternatives to jet propulsion. Jumping to earth from these balloons is an extreme sport.

Harnessing gas to fly into the stratosphere seems impossible to resist. We are taunted daily by birds who inhabit the air, defying gravity with apparent ease. Living every day with the vast expanse of the sky above, we have come to associate it with freedom, wonder and aspiration. Falling is easy, ascending is hard; this polarity is embedded in the geography of our dreams and expressed in our values. Success in life is associated with going 'up' in the world. We continue to invent balloons and other flying machines to do just that.

9. Sublime

I remember Felix Baumgartner jumping from a high-altitude balloon thirty-nine kilometres above the Earth's surface. It was streamed live on the internet in 2012. Before he jumped, we saw footage of Felix in his capsule, on the edge of space preparing to leap. Below, we could see the blue planet Earth in all its spherical magnificence. Felix was wearing a spacesuit because his balloon had reached the stratosphere. At that height there is very little of the Earth's gas atmosphere and almost no oxygen. The temperature outside the capsule was −57°C. As I waited, watching the live video feed, I envied him being up there between heaven and earth. This place where the gas atmosphere of our planet ends and the mysterious sublime state of nothing stretches out into the universe.

The materiality of space has puzzled humans throughout the ages. What really is it? Surely space can't actually be nothing? The ancients agreed. Aristotle declared that 'nature abhors a vaccum'. The heavens were thought to be filled with a sacred material. The Greeks called it 'aether', the substance the gods breathed, the fifth element, separate from the four that made up the earthly realm: earth, air, fire and water. It allowed light from the stars to propagate, and by medieval times it was also holding planets in their orbits. Even when Newton proposed gravity as a force in 1666, it relied on aether to propagate across the solar system. But no one could actually find a trace of this material, and as science began to rely on it more, so finding it became more urgent. The story of the search for this material starts back on Earth: the same Earth that Felix Baumgartner hurtled towards at 844 mph as he jumped from his balloon on 14 October 2012, almost certainly not thinking that the technology protecting him from the vacuum of space was in any way linked to this ancient quest for aether.

An Italian and a student of Galileo called Evangelista Torricelli was one of the first to make a breakthrough in the search for aether in 1641. His experiment was simple and elegant. He took a tube of mercury and turned it upside down in a bowl of mercury. Remarkably such an experiment shows that mercury does not rush down to the bottom of the tube pulled by gravity as you might expect. It falls a short distance and then stops. For a metre column of mercury, roughly 76 centimetres of it stay up the tube

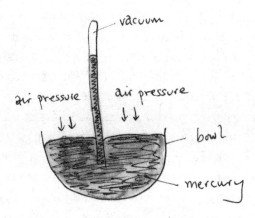

Evangelista Torricelli's first experiment to
understand the formation of vacuums

defying gravity. But there is a gap at the top where 24 centimetres of mercury used to be but are not there any more.
Torricelli asked what is in the gap. It is not air, because no
air could get in. So it is a vacuum, just like the vacuum of
space, and presumably filled with aether. Could this invisible aether be responsible for the mysterious force holding
up the mercury against the force of gravity?

The answer is no. Torricelli showed that there was a
much simpler explanation. The air we breathe forms
atmosphere on our planet, and despite being a gas it has
weight. It pushes down on us and everything it surrounds.
It is this air pressure that pushes down on the bowl of
mercury pushing the mercury up the tube. At the same
time the column of mercury is pulled down by gravity.
When those two forces are equal determines the height of
the mercury. This is why the height of the column changes
depending where you are on the Earth's surface. At sea

level the height of the mercury column is 760 millimetres. If you go up a mountain, the column gets smaller. This is because there is less air above you, less air pushing down on the surface of the mercury in the bowl, so less pressure pushing the mercury up against the force of gravity. If Felix Baumgartner had done the experiment in his balloon thirty-nine kilometres above sea level, he would have found that the tiny amount of atmosphere above him pushed so feebly that the height of the mercury column would have been 3 millimetres. So the vacuum doesn't do anything, its role is not to push back. What Torricelli had done was to find a way to create a vacuum on Earth. It had lots of technological implications, one of which would end up being the creation of TVs, computers and vacuum cleaners. But before that a more immediate invention beckoned, a way to measure atmospheric pressure: the barometer.

The barometer turned out to be able to predict the weather. Or at least some aspects of the weather. It could detect invisible changes in air pressure associated with different weather patterns, because they changed the height of the column of mercury in the glass tube. Those analysing weather patterns realized that high pressure was often associated with clear skies and sunny weather. While low pressure (a small column of mercury) accurately preceded rain and storms. The phrase 'the mercury is sinking' started to become used by sailors. It meant that the height of the column of mercury in their barometers was decreasing, indicating a low-pressure weather system was approaching and potentially a storm. Now they didn't have to pray to the wind gods or leave them offerings in order to know when was a good time to set sail. To this day air pressure is still measured in millimetres of mercury, denoted 'mmHg' as a result of this invention 400 years ago.

The development of this weather-forecasting tool was an unexpected bonus of exploring nothingness, but medieval scholars still had the puzzle of vacuums. Surely a region of the glass barometer with absolutely nothing in it was impossible: it had to be filled with something, even if that something was not air. Light travelled through the space at the top of a barometer, just like it travelled through outer space. So, they argued, they both should be filled with aether. They considered it a fundamental element of the universe, a perfect substance, but one that could perhaps be chemically isolated.

So the quest to isolate and distil aether began. It was led by the alchemists, who called it quintessence (the fifth element) and thought it could be used as a medicine to cure disease. Illness at the time was thought to be something that came from within a person, an imbalance of the four humours: black bile, yellow bile, blood and phlegm. Quintessence, the perfect substance, it was argued, could balance these humours and thus cure a person of illness. Others came to believe that quintessence was the fabled philosopher's stone that could turn base metals into gold. Once again it was a question of balance: lead had an imperfect balance of the fundamental substances sulphur and mercury and was thus a base metal that was soft and corroded easily. Quintessence could adjust the balance and so make this substance into perfect gold. Success in distilling quintessence would bring fame and wealth, but more importantly complete their quest to become close to God by studying and understanding God's Creation. And so the search for quintessence became a holy quest.

The person who made the next big breakthrough was not an alchemist though, but the mayor of the German town of Magdeburg, Otto von Guericke. As a politician he travelled across Europe, and this meant he was exposed to new ideas and the big scientific problems of the day. A devout man, he got to hear that quintessence might be the substance that filled the vacuum at the top of a barometer. Not being an alchemist turned out to be an advantage, because he did not have preconceived ideas of the right way to obtain quintessence. While alchemists were using all sorts of methods of chemical distillation, Otto did something completely different: he decided to isolate nothingness mechanically.

To do this he invented an air pump. It is a device we would recognize today as similar to a bicycle pump, except that the valves are reversed, so that each stroke of the cylinder removes air from whatever it is connected to, and then on the return stroke prevents the air from coming back. The mechanism is simple, but the execution is not. Whenever you remove air from a container, the air pressure outside the vessel creates a force driving air back into the container. This force gets bigger the more air you remove. Any leak in the valves or the fabric of the container destroys the vacuum. So to make it work requires precision engineering.

We take the accuracy and intricacy of screws, gaskets and valves for granted today. In the seventeenth century

Public demonstration of the Magdeburg Spheres

such precision engineering was just beginning: for instance, the mechanical clocks in city centres were only able to keep time to an accuracy of ten minutes in a day. Nevertheless, through ingenuity, perseverance and many failures, Otto succeeded in constructing an airtight pump. Despite this engineering success he probably wouldn't have been credited as being pivotal to the understanding of vacuums if he hadn't also been a bit of a showman. He showed the power of his air pump with a demonstration that would blow the minds of everyone who saw it.

Otto made two hemispheres of bronze which were machined so accurately that when they were placed together, they fitted to each other exactly. One had a small pipe incorporated to allow Otto's vacuum pump to be fitted. Then he assembled the important people in the land, including the King, to witness something incredible. He showed everyone the two hemispheres. They were just two pieces of not very interesting metal. Then he put them together to create a hollow sphere of metal. Next he used his air pump to remove the air from this internal space and create a vacuum. Now there was nothing physical holding the hemispheres together. No bolts, no straps, no welding, no glue. Everyone could see that. Nothing. Then he assembled two teams of eight horses. The first team of horses was harnessed to one half of the now joined Magdeburg Sphere and the second team to the other, the two teams facing in opposite directions. Presumably the horses neighed and stamped their hooves, not knowing what was going on. Perhaps the wind dramatically ruffled their manes. Then Otto drove the two teams away from each other, trying to make them pull the two halves of the sphere apart. They pulled against suction. But they could not defeat it. A pump, and some precision engineering, had created a suction that could defy sixteen horsepower. But it wasn't a force from the vacuum inside. Just as with the barometer, atmospheric pressure was pushing the two hemispheres together, and, without air inside, nothing was pushing back.

Soon engineers and instrument makers across Europe were building their own air pumps and using them to explore the anatomy and properties of vacuums. As with Otto von Guericke's demonstration, part of the magic was the public nature of the experiments. Famous scientists of the day such as Robert Boyle started using air pumps to evacuate glass vessels, so that anyone who cared to look could see what was going on inside. These demonstrations became public entertainment as well as pushing forward the science.

Does a bell ring in a vacuum? Answer, no: sound waves need air as a medium to travel. Does a candle burn in a vacuum? Answer, no: but oxygen had not been discovered yet, so there was no good explanation. Can an insect fly in a vacuum? Answer, no: wings need a gas to create lift. Can a snail survive in a vacuum? Answer, no: it dies. Can a mouse survive in a vacuum? Answer, no: it dies. Can a bird fly in a vacuum? Answer, no: it flutters and then dies in agony. What happens if you put a compass in a vacuum: does it still point north? Answer, yes: magnetism is unaffected by a vacuum. Does electricity flow in a vacuum? Answer, yes: and light travels through it without a hitch too. Ah ha – you're thinking, a clue! And yes, you're right, this is exactly why the scientists of the day were so excited about these discoveries.

So it was that Otto von Guericke's air pump was crucial to build the evidence that although some things, like sound, needed the medium of air to travel, others, such as light, magnetism and electricity, did not. Perhaps they were special in some way, or perhaps they were connected to whatever there was in a vacuum that allowed them to travel, not just through a vacuum but across space and time. The potential role of quintessence was expanding.

In 1768 the spectacle of the popular and mysterious air pump experiments was captured in a painting by Joseph Wright of Derby. Called *An Experiment on a Bird in the Air*

An Experiment on a Bird in the Air Pump, Joseph Wright of Derby

Pump, it now hangs in the National Gallery in London. There are wonder and sorrow in that painting. The central figure conducting the experiments is a man looking out towards the viewer with an impartial expression, as if to say, 'This is how to understand the world.' Some of the onlookers are covering their eyes, distressed at the cruelty of experimenting on live animals. Others are staring intently at the demonstration, utterly fascinated by this insight into how the universe works. On the table is a set of Magdeburg Spheres, a reference to the origin of these pumps and the quest to understand air, vacuums and quintessence.

I wish I could say that this was one of the paintings I remember as a kid. I wish I could say that I stood transfixed in front of this painting on one of our many visits to the National Gallery, where my mum's relationship with the gods of parking allowed us to access the museum with ease. But unfortunately I don't remember seeing this painting as a child, even though Mum almost certainly would have shown it to us. Yes, because it is a masterpiece, but also because it's a tangible connection between her and Dad. He was a renowned metallurgist and very much involved in exploring how the world works through experiment and philosophy. She would have appreciated the mystical and ceremonial quality of the painting, with the candle-lit setting in particular lending the scene a spiritual air; this was all lost on us boys, who were probably running amok in the gallery. I was perhaps like the boy in the painting who is not looking at the experiment but instead fiddling with the window blind, and in doing so letting moonlight into the room. This is an intentional reference by the painter to the Lunar Society and the questions being asked at the time about how light travels from the Moon to Earth through space. One of the reasons why something like quintessence had to exist was because light waves needed a medium by which to travel through space. Sound waves travelled through air, sea waves travelled on water – what was the equivalent medium for space? Scientists called it luminescent aether – renamed because they couldn't find quintessence.

Meanwhile the engineers, who had been spending a lot of time making vacuums in glass containers, were getting annoyed at having to continuously pump out the container every time they wanted to do an experiment. What if, they reasoned, once the glass vessel contained a high-quality vacuum, the glass was melted to seal the vacuum inside the chamber. This produced a permanent vacuum inside the glass on which to experiment. Of course, you could not move things in and out of the container once it was sealed, so you had to decide what you were going to experiment on and leave it in there. Metal wires could be used, for instance, connected at either end of a glass tube, or glass bulb as it was called. When a voltage was applied, electricity would flow through the tube and the wires would grow very hot. This caused them to glow red hot. It was the birth of the electric light bulb, an invention deemed so ingenious that the universal symbol for having a brilliant idea is a light bulb.

Early versions in the 1800s emitted light only for a short time, after which the hot glowing wires, called filaments, would then break. Scientists realized that for electric light bulbs to replace candles or gas lamps, the electricity would need to heat up the filament to temperatures exceeding 1,500°C. But there was a problem: this temperature exceeds the melting point of most of the metals used to conduct electricity. By the time Humphry Davy had a go in 1802, the metal platinum was the leading contender, with a melting point of 1,768°C. But white-hot platinum vaporized at that temperature quite quickly and so the filaments didn't last long. They were also very expensive. A cheaper conducting material was needed with a high melting point. The British chemist and inventor Joseph Swan used graphite, which seemed perfect because solid carbon doesn't melt at all. You have to increase the temperature to 3,642°C before it gets so hot it evaporates into a gas, a process called sublimation. Swan took out a patent in 1860 and that should have been the beginning of a bright future. But the difficulty was that carbon reacts very easily with oxygen in the air. Of course, the vacuum inside the bulb should have meant that there was no oxygen. But the early mechanical air pumps did not produce perfect vacuums. They could reduce the air pressure enough to suffocate a bird, kill a mouse, or prevent an insect from flying, but there was still a small amount of air left in the glass bulbs. The oxygen in that air reacted with the carbon filaments and that destroyed Joseph Swan's early electric light bulb.

It was only by 1875 that vacuum technology improved enough to create an electric light bulb with a carbon filament that could glow white hot in a vacuum, providing ample light for forty hours. Swan started with his own house in Gateshead, then lit a whole street in Newcastle-upon-Tyne, and then the Savoy Theatre in London. It was the future. The American Thomas Edison is often credited with inventing the light bulb. He didn't. What he did was to see that the future of lighting was electric. He perfected the production and marketing of lighting systems, including the bulbs. He is famous for stating that an idea is only a small part of invention: 'What it boils down to is one per cent inspiration and ninety-nine per cent perspiration.'

The perspiration in the case of the electric light bulb was the enormous number of experiments Edison performed on different designs of bulb. Most of them failed. But proof of the importance of perspiration and systematic testing was that one of his experimental light bulbs,

The many types of vacuum tube that came into existence as the
electronics industry blossomed in the early twentieth century

which seemed to have no use, turned out to be the beginning of computers.

In essence it was just a light bulb with a broken filament.
What Edison's engineering team noticed was that you could
still get electricity to flow through the vacuum, but only if
the filament was hot. The electrons would jump across the
gap between the broken filament from the negatively
charged end to the positively charged end but not the other
way. This was the birth of a component that would kickstart the electronics industry, and it was called a vacuum
tube. These vacuum tubes acted as valves, the equivalent of

Early TVs resembled laboratory equipment

the taps in your kitchen that control water flow. These valves allowed electric signals to be turned on and off by another electric signal (which heated the filament). This was a pro-grammable tap that could tune and amplify electricity. It led to the development of the loudspeaker, the radio and the TV, the last-mentioned having at its heart one giant vacuum tube, called a cathode-ray tube.

Cathode-ray tubes have Edison's hot filaments at one end and high voltage at the other, where there is a screen. The 'cathode-ray' is not a ray of light but a ray of electri-city. It literally flies across the vacuum tube, but the only reason it reaches the screen is because there is no gas in the way for the ray of electrons to bump into. When the electricity hits the screen, it lights up because of a special coating called a phosphor. Now there is a bright spot on the screen. To make these TVs work, the ray is scanned across the screen very fast, row by row, so that each part of the screen is hit by the electricity twenty-five times every second. You would observe this scanning dot if you could see that fast, but you can't, so instead you see a continuous image of a wizard casting a spell, or a tornado transporting a house through the air. I still remember these TVs from my childhood: we watched films like *The Wizard of Oz* on them. They resembled vacuum laboratory equipment because that's exactly what they were. When you turned the TV off there was a click and the screen suddenly went blank, except for a single dot in the middle. This dot was the place where the last electrons had hit the screen. The place still glowed for a second before fading to nothing. It was always a sad moment for me and my brothers. The appearance of that dot meant we were going to bed.

TVs in those days were huge heavy things. They were weighty because the cathode-ray tube was made of glass, and it was not just ordinary glass. A by-product of accelerating electricity to create that dot on the screen is the creation of x-rays, the same x-rays that are used in hospitals to detect broken bones and cancer tumours (and yes, hospital x-ray machines are also vacuum tubes). To protect TV viewers these x-rays had to be stopped before they escaped from the vacuum tube and radiated everyone watching the TV programmes. That meant adding lead to the glass, which absorbed the x-rays. This worked, but lead, being a very heavy element, increased the weight of the TVs, which were the size of armchairs in my childhood.

Most of these enormous TVs are gone now, freeing up a lot of space in our living rooms but leaving a feeling of nostalgia for the simplicity of when we only had three TV channels to watch. They have been replaced by liquid crystal flat-screen technology controlled by silicon chips, with hundreds of TV channels. This materials science invention from the 1950s created the revolution in computing, replacing glass vacuum tubes. Silicon chips are a core technology in every computer, in every mobile phone, in every car, washing machine and hospital. These silicon chips need to be manufactured in ultra-high vacuums: otherwise they become contaminated with impurities from the air which render the chip worthless. Thus a thousand-year-old quest to create the purest 'nothing' still continues. And there is still plenty to do, since we as yet can't even make a vacuum as pure as that found in outer space, which is millions of times purer.

For most people the holy grail of vacuum technology is not their mobile phone, despite its importance and much as they might love it. It's not the vacuum used in medical technology to produce x-rays, much as they care about its importance for diagnosing illness and tooth decay. It's not the vacuums used in the scientific equipment in every lab in the world without which scientific research would come to a standstill. These are all too remote and hidden from view to be of daily concern to citizens of the world. No, for most people the most important vacuum in their life is inside their vacuum cleaner. These machines, like the early steam engines, harness atmospheric air pressure created by the hundred kilometres of air above our heads to clean our homes. They create a vacuum inside the machine which causes air to rush in to equalize the pressure, and in doing so it sucks up dust as the vacuum cleaner kisses the floor. It is so simple, and yet so marvellous. It has made all of our homes less filthy, especially homes with fitted carpets, which would otherwise be dirty, dusty and smelly. The vacuum cleaner is the stalwart of the home, creating order and cleanliness. It has even played its part in creating more equality between the sexes, making cleaning faster and more effective – freeing time for other things, like careers and hobbies – and also lowering the barriers to those reluctant to contribute to cleaning the home.

This brings us back to the search for luminescent aether, the perfect substance, said to inhabit space. By 1905 Einstein's Special Theory of Relativity banished the need for aether to explain how gravity works and how light travels through space. According to this theory, there is no need for aether and 'nothing' really does exist. It is the creation of nothing inside a vacuum cleaner that harnesses atmospheric air pressure to clean our homes. It's the nothing inside a light bulb that allows light to emerge. It's the nothing inside an x-ray tube that helps doctors diagnose illness. It's the nothing in vacuum chambers that allows us to test the safety of space suits, enabling Felix Baumgartner to safely jump from a balloon on the edge of space. The purer the nothing, the more effective it is. Less is quite literally more. Especially, as we find out in the next chapter, when small discrepancies in the nothingness reveal hidden truths.

10. Sprites

Nothing reinforces my feeling of being an oddball more than attending a black-tie dinner. I arrive at these dinners improperly dressed and am greeted with a concerned look by my hosts. Their eyebrows note my lack of a black bow tie and find instead an open-necked flowery shirt. They say things like, 'Can I help you, sir?' when what they mean is, 'What the **** are you wearing!' Sometimes they give me a tie to wear and insist I put it on.

Why do I do it? This is the question I ask myself now that my mum is dead. She asked it a lot before she died and

I had no really good answer. Conform, fit in, wear polo shirts, this is all she wanted for me, despite the fact that she herself was an eccentric woman. Although, to be fair, she wasn't eccentric in her attire. Perhaps that is what she learned in life: conforming in dress paradoxically gives you more freedom. If it is true, it's a lesson I never learned, along with the many other things I rebelled against. Having a rebellious nature does have some advantages in science though, and there is no better example of this than the quest to find the noble gases, a family of invisible gases that blew apart scientific orthodoxy.

The story starts in the late nineteenth century when a scientist called Lord Rayleigh was obsessing about the known chemical elements. He noted that the weights of the atoms displayed a pattern. This pattern, depicted in the Periodic Table of Elements, showed that there were families of elements that had similar properties. Most of the weights of the elements seemed to be multiples of the smallest atom hydrogen, indicating perhaps that the hydrogen atom was the basic unit of all atoms. And yet there were some anomalies. For instance, lithium atoms are almost exactly seven times the weight of hydrogen, but not quite – they are a tiny bit more than that. The weights of most of the heavier elements were not neat multiples of hydrogen either. Lord Rayleigh didn't like this. The universe wasn't a free-for-all: he thought that the elements should fit a system, and he decided to make them fit.

To iron out the problem, Rayleigh decided to weigh them again. Calculating the weight of gases is, as you might expect, not very easy. Rayleigh had to create a vacuum in a glass flask, weigh it to get a measurement for the flask without gas, then add pure hydrogen and weigh it again. It was tricky, not just because achieving a pure vacuum was difficult, but also because the buoyancy of the air in the room affected the result. Imagine trying to weigh yourself while at the bottom of the sea: the density of the water around you gives you buoyancy and so makes you lighter. This is true of the air around you too: the amount of buoyancy depends on the density of the air. This in turn is affected by the temperature and humidity in the room – they vary from day to day. Instruments for measuring these factors were available to Rayleigh, such as the barometer and the thermometer, but it was exacting work and it took him years to achieve reliable results. You might wonder why he persisted, but it's a mark of the rebellious nature of scientists like Lord Rayleigh that they won't believe data or theories that look wrong (to them). It is an instinct that drives scientists to work obsessively on problems for years, even decades.

Having weighed hydrogen to his satisfaction, Rayleigh moved on to nitrogen. This gas, which is left over from the formation of the Earth, makes up 78 per cent of the air we breathe. This gas doesn't chemically react with much, and so it seemed obvious to Rayleigh that the way to produce pure nitrogen was to remove the oxygen, water vapour and carbon dioxide in air. He did this using various chemical procedures, and then he measured the weight of the remaining gas to get an accurate value for the weight of nitrogen atoms. He wasn't quite satisfied with that answer, or maybe there was a nagging doubt in his mind – in any case he decided to double-check the result by finding another way to end up with pure nitrogen. This time he took ammonia gas and reacted it with oxygen. This chemical reaction produces water and pure nitrogen gas. He did this and measured the gas to make sure that the weight of the nitrogen atoms in ammonia was the same as in air. But it wasn't. How could this be? He had made a mistake of course. Or, at least, that was what he assumed, and so he did the experiment again, refining it for another two years. But after checking and double-checking he always found the same result, which was that the two weights of nitrogen atoms were different. This irked Rayleigh. He came to the annoying conclusion that the extra weight of nitrogen when it was in air was not extra weight at all – it was another gas. This invisible gas was hiding in the air. In experiments with what he assumed was this new gas he found it impossible to make it chemically react with anything else, even the most reactive substances such as fluorine and lithium. This was extra annoying, and because of this trait the gas was named 'argon' – which is from the Greek meaning 'inactive' or 'lazy'. Even more annoyingly the new gas didn't fit into the Periodic Table of Elements.

The Periodic Table of Elements is the chemists' equivalent of the map of the universe. It tells you the names and symbols of all the known chemical elements from which everything is made, of which 118 are currently known to exist. It starts with hydrogen, the smallest and lightest atom, and lists all the others up to element number 118, oganesson. The word 'Periodic' refers to the observation that some elements seem similar to each other in terms of

Reihen	Gruppo I. — R^2O	Gruppo II. — RO	Gruppo III. — R^2O^3	Gruppo IV. RH^4 RO^2	Gruppo V. RH^3 R^2O^5	Gruppo VI. RH^2 RO^3	Gruppo VII. RH R^2O^7	Gruppo VIII. — RO^4
1	H=1							
2	Li=7	Be=9,4	B=11	C=12	N=14	O=16	F=19	
3	Na=23	Mg=24	Al=27,3	Si=28	P=31	S=32	Cl=35,5	
4	K=39	Ca=40	—=44	Ti=48	V=51	Cr=52	Mn=55	Fe=56, Co=59, Ni=59, Cu=63.
5	(Cu=63)	Zn=65	—=68	—=72	As=75	Se=78	Br=80	
6	Rb=85	Sr=87	?Yt=88	Zr=90	Nb=94	Mo=96	—=100	Ru=104, Rh=104, Pd=106, Ag=108.
7	(Ag=108)	Cd=112	In=113	Sn=118	Sb=122	Te=125	J=127	
8	Cs=133	Ba=137	?Di=138	?Ce=140	—	—	—	— — — —
9	(—)	—	—	—	—	—		
10	—	—	?Er=178	?La=180	Ta=182	W=184	—	Os=195, Ir=197, Pt=198, Au=199.
11	(Au=199)	Hg=200	Tl=204	Pb=207	Bi=208	—	—	
12	—	—	—	Th=231	—	U=240		— — — —

The Periodic Table compiled by Dmitri Mendeleev (1869)

their chemical properties, despite having very different weights. For instance, lithium, sodium and potassium are all soft metals that react violently with water. Each one is heavier than the last, but because their weights increase by a similar amount, you can think of them as a big atom, a medium atom and a small atom of the same type. Hence the idea that they are a family. This feature was noticed by a Russian chemist called Dmitri Mendeleev, who assembled one of the first versions of the table in 1869 and became famous for doing so. At the time only sixty-three elements were known to exist, but he spotted different families of elements and put them in columns. He also noticed that there were some gaps in the table where an element should exist but had not been discovered. For instance, there was a gap in the family of boron and aluminium that Mendeleev's table predicted should have an extra family member with an atomic mass of 68. Six years later a new metal similar to aluminium but softer was discovered, named gallium. It fitted the spot. This was a big win for Mendeleev's Periodic Table.

Other new elements followed that filled in the holes in the Periodic Table, and all was going swimmingly until Rayleigh's new gas, argon, came along. There was no space for it – it did not fit in the table and it was not welcomed by the scientific world. It is easy to believe that a metal like copper exists, because you can hold a lump of it in your hand. In contrast most gases are invisible and intangible, and so our belief in their existence is related to less direct evidence. Still, most people don't doubt that oxygen and hydrogen exist, since this fits their world view. In the case of argon, most chemists at the time were sceptical of the evidence, especially because it did not fit the mighty Periodic Table. Only one chemist, William Ramsay, a professor of chemistry at my own institution, University College London, took the trouble to do the experiments himself. He found the same results as Rayleigh. They joined forces and took the brave decision to announce in 1894 to the British Association that argon was a new gas. It was colourless and odourless, it did not react with anything, it did not fit into the Periodic Table, and that although this was a kind of heresy, it was true. The audience were appalled and sad in equal measure. It was clearly an experimental error. These errors are common in science, but the audience expected better of these two excellent scientists.

The importance of argon not fitting into the Periodic

Table cannot be overstated. At the end of the nineteenth century, chemistry was undergoing a revolution with new elements being discovered every few years. Everyone who had the skills to look for new elements was doing so. Fame and money beckoned. At the height of Periodic Table mania, not a month went by without someone claiming to have discovered a new element. Almost all of these claims were dismissed as wrong or fraudulent, and it was assumed that argon was another one of these. The fact that argon didn't react with anything was really odd. It is almost the definition of an atom that it forms bonds with other atoms. Being unreactive made no sense.

After Ramsay's and Rayleigh's proclamation about argon gas, the British Museum alerted Ramsay to a uranium mineral called cleveite from Norway. It had a strange property: when heated it gave off a gas that wouldn't react with anything. Excitedly Ramsay got hold of some cleveite and repeated the experiment. A gas bubbled off. He weighed the gas but found it couldn't be argon because it was too light. To double-check, he sent the sample of this new gas in a glass tube to his colleague William Crookes, who was the world expert on spectroscopy: a key technique in the armoury of chemists to identify chemical elements, especially ones that are invisible.

The story of spectroscopy starts with Isaac Newton. He took a glass prism and shone a ray of white light from the sun through it. The prism created a range of colours from dark blue to green through yellow, orange and red. He called this a spectrum of light. He showed that these colours were not created in the prism but were created in the sun and travelled together as a mixture – it is this mixture that gives the sensation of white light. His convincing evidence was that recombining the spectrum of light produced a beam of white light. This was big news back in the seventeenth century and also provided the explanation for how colourful rainbows are created from sunlight: it is the drops of water in the sky that behave like prisms and split the sunlight into its constituent colours.

But if you look carefully at the spectrum of sunlight, you see that there are some gaps in the colours: these are black lines that mark the absence of a particular wavelength of light. Called spectral lines they are due to the absorption of this wavelength of light by the elements inside the sun. Two scientists, Gustav Kirchhoff and Robert Bunsen (of Bunsen burner fame), worked this out by heating pure samples of known elements on Earth until they glowed and matching them with spectral lines. So we know that the sun is mostly hydrogen from the spectral lines, not because anyone has ever visited the sun and taken gas samples. The sun also has small amounts of all the other known elements such as carbon, nitrogen and oxygen: again we know this from the spectral lines. But sunlight also has some spectral lines that don't match any of these elements. When these lines were discovered in 1868, it was speculated that this might be the signature of a new element and it was given a name, helium, from Greek *helios*, meaning the sun. (Spoiler alert: as we have seen and as everyone knows, helium exists!)

It was this spectroscopy technique Ramsay wanted to use on his samples of the new gas that bubbled out of the cleveite rock. He already had the spectral lines of argon, so

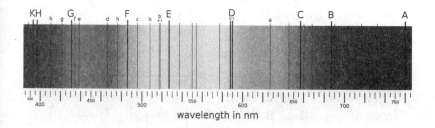

Solar spectrum from the sun showing spectral
lines associated with hydrogen in the sun

would this new gas turn out to have the same ones? The answer was no. This gas was not argon. It wasn't nitrogen either. It wasn't anything anyone had seen before on Earth. But they did match the spectral lines of the mystery element in the sun, helium. This was another headache for Ramsay: he didn't really want to be associated with crank scientists claiming that the sun contained a special element. Also, if this was a new gas, there was nowhere on the Periodic Table for it to exist. It was another misfit. Many chemists attacked him for being a slap-dash scientist like so many others out there, quick to claim amazing new results.

Could it be that the all-conquering Periodic Table was wrong? Or at least missing a whole column, a place where argon and helium, if they existed, would belong. But if there was to be a new family of gases there would need to be more than two gases in it. So far Ramsay had an extremely light one and a medium heavy one. He guessed there should be another gas with a weight between these two. He could even estimate the approximate weight of this new gas, since the atomic weight of helium was 4 and argon was 40, so he was looking for an invisible, odourless, unreactive gas with a weight of around 20 atomic units. Where to look though? Ramsay remarked at the time: 'here is a supposed gas, endowed with no doubt negative properties, and the whole world to find it'. But even that is an understatement: even if this gas existed in the universe, there was no reason to believe the gas was even present on Earth.

Years went by with Ramsay and his assistant, Morris Travers, dissolving rocks similar to those where he found helium, in the hope that another slightly heavier but inert gas might bubble out of them. They got nowhere. Then at some point they realized that temperature was the key to the problem of finding more inert gases. They reasoned as follows. If a gas is unreactive, then when the Earth formed it would have bubbled up from the core into the atmosphere. This might have taken billions of years but once in the atmosphere it would stay there because it can't react with anything. Thus the place to look for a third unreactive gas was not in rocks but in the air.

Ramsay and Travers took air and removed the major constituents: oxygen, nitrogen, water and carbon dioxide. They then cooled down the remaining air, which condensed into a cold liquid. Then, the clever bit: by slowly heating up this liquid, the different gases, each with a different boiling point, would boil out of the liquid one by one, allowing them to be captured. The lightest gas, which was helium, boiled off first, and the next one should have been argon, unless there existed another one – a trace element to fit in the gap in the family of gases between helium and argon. But was it there?

After many months of experiments Ramsay and Travers had a glass tube filled with an inert gas that was lighter than argon but heavier than helium. They then passed electricity through it to look for spectral lines but needn't have bothered with special equipment, because this gas turned out to be the show-off of the family. The tube lit up bright crimson red. They had discovered neon.

Ramsay and Travers also discovered other unreactive gases, all hiding in the air we breathe. They named one krypton, the Greek for 'hidden', and another xenon, the Greek for 'stranger'. This family of gases (called the 'noble', meaning unreactive, gases) are all invisible, odourless and unreactive, but they are there. The quantities in the air we breathe are truly tiny, except for argon, which is two hundred times more abundant than carbon dioxide in the Earth's atmosphere.

Ramsay was awarded the Nobel Prize for Chemistry in 1904 for the discovery of the noble gases. It was a triumph for him, especially because he had had to put up with constant criticism about whether he was delusional or just a poor experimenter. Ultimately he was vindicated because others could and did repeat his experiments and found the gases too. 'Seeing is believing', or in the case of invisible gases, 'measuring the spectral lines is believing'.

The reason for noble gases being unreactive is that the electrons inside them are all paired up and reluctant to break that pairing to interact with the electrons of any other atom. Hence they don't react or bond with most things, which makes them perhaps the most mysterious of all chemical elements. They don't even like to bond to themselves to form a liquid or a solid, which is why they are gases at normal temperatures. So krypton, which is heavier than iron, is a gas. Similarly radon, which is heavier than gold, is a gas. The public was quite impressed with this, but the question remained: what use were these gases? They didn't react with anything, and so didn't form the basis for new advanced materials. As we have seen, helium was useful in airships, but neon, argon, krypton and xenon didn't seem to solve any of humanity's pressing problems. They might well have disappeared back into obscurity had modernity not beckoned.

A French inventor, Georges Claude, was the first to propose a use by considering whether the bright red glow of neon gas might be used as a light source. In the early twentieth century, electric light bulbs were already popular, but they burned out quickly. This was because they worked by passing an electric current through a thin filament of carbon, which glowed white hot. The high temperature produced incandescence that illuminated a room and gave out heat. Due to the high temperatures these delicate pieces of carbon filament often broke. The search for better materials eventually produced tungsten filaments, but they still had a limited lifespan. In contrast a glass tube of neon needed no filament to light up: all it required was neon gas, and even that wasn't used up or chemically changed in any way by passing electricity through it. Neon lights potentially had an infinite life.

The problem was aesthetic – the colour was electric red, which made everything look weird indoors. Using neon light in the kitchen changes the colours of food, so green beans look black, and potatoes look red. In the living room wooden furniture looks grey and it was useless trying to read a magazine using this light. For street lighting, neon lights were equally hopeless: too inefficient in energy use to light large areas – except for one role, advertising. Because the glass tube could be formed using glass-blowing techniques, the neon light could be in any shape. It could even spell out a word. And so it was that at the turn of the twentieth century the word 'CINZANO' beamed out over the Champs-Élysées, advertising a popular vermouth to passing Parisians. The neon sign could be seen clearly at night even if the weather was foggy. It was an urban beacon guiding the lost and weary to a rather refreshing drink.

Soon the neon sign became the signpost to a future that was about to abandon old ways of doing things. The automobile, radio and TV all followed in quick succession, transforming everyone's lives in a dizzyingly short space of time. Art reflected the technological shift through the movements of cubism and futurism, and a new architecture in the form of art deco. Neon may not have fitted into Mendeleev's Periodic Table, but its crimson glow did fit right into modernity: it was not subtle, nor was it sophisticated, but it was futuristic and hopeful, and perfectly captured the mid-twentieth-century zeitgeist.

Neon was loved most in the USA, a country that had no regrets about abandoning the past and embracing the new. The streets of Manhattan became packed with neon signs advertising bars. It was the ultimate aperitif. The neon signs outside were a taste of the spectacle and excitement to be had inside, whetting customers' appetites for dancing,

Las Vegas neon sign

drinking and music. No night club was complete without one. The bigger the sign, the better, and that went for advertising products too. Soon Times Square in New York had the highest concentration of the element neon in the known universe, magnifying the attraction of the Broadway theatres and night life in the bars. Neon was new, it was fun, and it spread across America, each light guiding the population into a new modern life. The ubiquitous neon 'OPEN' sign outside a motel conveyed to the weary traveller more than information about the availability of accommodation. Even the most drab of motels became instantly contemporary and space age with a neon sign outside.

By the time Las Vegas defined itself as the world capital of gambling, neon light manufacturers were using other noble gases to make different colours, but the generic name 'neon light' for a tube light had stuck. Argon emits a

Soviet neon signs, 1966

violet light, which can be made more vibrant by adding mercury vapour. Helium emits a pinkish-red one, xenon is deep lavender and krypton has a white-yellowish glow. This kaleidoscope of flashing bright lights became synonymous with Western good-time capitalists.

Not that the communists entirely ignored neon. After Stalin died in 1953, there was a period when Khrushchev encouraged the Soviet cities to match the glamour of Western capitals. This 'neonization' of communism resulted in a flourishing of signs not as commercial advertisement but as a different type of propaganda. The adoption of neon was a deliberate attempt by the Soviet authorities to quell the disquiet and dissatisfaction with perceived differences in lifestyle of Soviet citizens comparing themselves to their counterparts in the West. Through snippets of TV, cinema and magazines, the Soviets had an insight into the kind of lives that those in the West were living, and they wanted a taste of modernity for themselves.

The glamorous status of neon might have surprised the Russian father of the Periodic Table, Dmitri Mendeleev, had he not been dead by then. It was the misfitting helium, argon, neon and krypton that broke his Periodic Table apart. The table had to be ripped up to create a new one that would accommodate this new family of gases.

During the twentieth century the noble gases rapidly became part of the life support system of modern economies around the world. As we have seen, helium became an important gas, not just for airships but also for cooling medical technologies such as MRI scanners and scientific instruments such as electron microscopes. Science would have stalled without these instruments. Neon, argon, xenon and krypton gases also turned out to have vital roles in our modern high-tech civilization, being essential for the lasers that make our telecommunication systems work. Argon lasers are used for miraculous eye surgery and helium–neon lasers are at work in barcode scanners and blood analysis.

Chemists, having transformed the Periodic Table once, did it again and again throughout the twentieth century to accommodate the many new elements that were discovered. To be honest, it makes the current version a bit messy, with different rows in the table such as the lanthanide elements

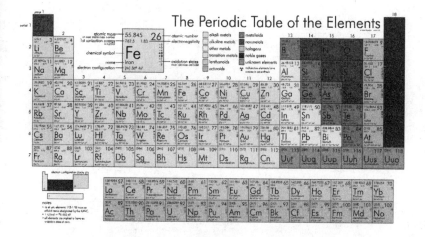

The current version of the Periodic Table showing
the family of noble gases in column VIIIA

(e.g. neodymium) and actinide elements (e.g. uranium)
having to be separated from the main body. This makes the
table an odd but iconic shape which is instantly recognizable. It is celebrated on the wall of every chemistry lab in
the world. The shape reminds us to honour the rebellious
spirit of Rayleigh and Ramsay, a spirit that is always in the
air, literally in the case of the noble gases – you inhale them
with every breath you take. You breathe them out too,
along with carbon dioxide. This is another gas whose invisibility has caused a great deal of argument and rebellion
against the status quo. These arguments are still raging, and
the established fossil fuel industry is holding out against
change, but, as we will see next, this can't last.

11. Phantom

I'm not good at prescribed relaxation. I hate taking my shoes off in public, and the phrase 'Get into pairs and hold hands' fills me with horror. Nevertheless, for my whole life I have tried to get better at relaxation and endeavoured to find yoga appealing. So in 1996 when a friend of mine invited me to a yoga retreat, I accepted.

Turning up at the converted barn in the middle of the Oxfordshire countryside, with the sun creating dappled shade, I thought, 'Yes, this is it!' Of course, it was so obvious, I should retreat from the world and take it easy in a beautiful place. No more scientific laboratories and complicated city life for me, I had found my path. I beamed at my friend.

When we got inside though, the sight of five or six young people in their bare feet and loose clothing made me uneasy. Due to my upbringing, when I had to go to a primary school run by a religious cult, I become anxious around people who meditate or use the word 'mindful'. Nevertheless, I changed into loose clothing, took off my socks and shoes, and stood awkwardly, longing for it all to end. I hoped my unease was not obvious, since I had once been chucked out of a yoga class for voicing my persistent inner thoughts out loud. I was 'creating a bad vibe', I was told, 'and we are overbooked anyway. Would you mind leaving?' So this time I kept my mouth shut.

We were welcomed, asked to find a space on the floor, to lie down on our backs, and to close our eyes. We were then asked to imagine our 'happy place', a place where we could return to in our mind and spend time when we were feeling stressed.

'Think about the sky, what is it like there?'

Sunny, I thought, blue skies.

'What can you see?'

A beach, I thought, with palm trees.

'What are you doing?'

I'm swimming, in the turquoise water.

'Who else is there?'

There is a beach-side bar shaded by palm leaves that serves cool beer . . . no, that's bad, isn't it? My happy place shouldn't need alcohol.

My worries about drinking too much beer then took me off into a series of other worries about wanting to live on a tropical island when I hadn't grown up there. Weren't all us Europeans just ruining these places? But tourism, that's important for the local economy, isn't it? Or is that a convenient fiction? On and on went my internal dialogue until I remember being wrenched out of it by the instructor saying, 'Everyone sit up and hang on to that happy feeling as we do some stretching.'

Since then, I've been to a lot of tropical islands, and I have to say that for whatever reason it has never created in me a feeling of relaxation. This is probably because I bring myself on the holiday and my intrusive thoughts dominate. These days I worry that the big threat to these islands is not tourism, but carbon dioxide. Or more precisely, the increase in carbon dioxide in the atmosphere which is causing sea levels to rise.

It was an American woman, Eunice Foote, who first flagged that carbon dioxide could cause climate change. She presented the theory at the American Association for the Advance of Science meeting in 1856. Scientists were initially quite positive about it, hoping that a better understanding and then manipulation of gases in the atmosphere might allow us to prevent new ice ages. Then in 1896 the Swedish scientist Svante Arrhenius showed that without an atmosphere Earth would be cold like Mars, with surface temperatures fluctuating from $-70°C$ to $20 °C$ between night and day. It is the atmosphere that stabilizes the temperatures on Earth by reflecting heat radiation (called infrared light) back to Earth rather than letting it disappear into space. It keeps us warm and is called the greenhouse effect, because greenhouses trap heat in a similar way.

The Earth's atmosphere is mostly made from nitrogen and oxygen, but these gases are not good at trapping heat, because they are transparent to this radiation: it just travels straight through these molecules. It is other constituents of the atmosphere such as water vapour, carbon dioxide and methane that are good at trapping heat. Water vapour is responsible for about 60 per cent of the Earth's greenhouse warming effect, but it is not in control of the Earth's thermostat because water vapour condenses into droplets forming clouds. Rain showers then remove it from the atmosphere. So it comes and goes from the sky. It is the gases that don't condense, such as carbon dioxide, that are in control of Earth's thermostat. Carbon dioxide makes up less than 1 per cent of the atmosphere but it has a huge effect on the climate, creating 25 per cent of the warming. This is because the carbon dioxide emitted from our cars, power stations, homes, aeroplanes and indeed our breath is an extremely efficient molecule for trapping heat.

In 1996 when I was failing to relax at a yoga retreat, the concentration of carbon dioxide in the atmosphere was 380 parts per million (ppm). That means there were 380 carbon dioxide molecules for every million molecules in the atmosphere. Now it is 420 ppm. The difference doesn't seem great, but it is this increase that has caused global warming over the past twenty-seven years. Carbon dioxide is invisible, and this invisibility makes it easy for us to ignore the increase in concentration. The temperature changes have been gradual, so it is easy to deny those too. But one change is very obvious: sea levels are rising continuously.

The water that causes rises in sea level has to come from somewhere – that seems obvious – and so we would expect the ice in the Arctic and Antarctic to be disappearing, which indeed it is. For the first time in history there is now a sea passage through the Arctic seas during the winter, when there used only to be solid ice. When climate scientists measure the ice coverage in the Antarctic, they find it is decreasing year on year. But the reduction of the ice is not enough to explain the sea level rises that we have experienced. The temperature of the oceans has increased too, and this means that the water in the oceans has got hotter and so has expanded, taking up more room and raising sea levels (this is responsible for approximately 40 per cent of the sea level rise).

Since 1900 the average sea level has risen by twenty centimetres. This doesn't seem too bad, the height of a tall drinking glass. No cities have yet disappeared, no countries have yet been submerged, but if sea levels keep rising, it is inevitable that this will be the fate of many of them. By the end of 2100, the conservative estimate of sea rise is 200–300 centimetres, but it could be much higher. Sea levels rose ten metres above present levels 125,000 years ago when the ice melted. The small low-lying islands will go first: the sun-drenched places with white sandy beaches I imagine when asked to envision my happy place in yoga retreats. We are on course to live through a period of history when we wipe these happy places off the map.

Situated in the middle of the Indian Ocean, the Maldives

Bathala island in the Maldives

is the lowest-lying country in the world. It comprises an archipelago of the most gloriously idyllic tropical islands you could ever wish to visit. But the people who currently live on the Maldives will probably never see their grandchildren grow up there, because the islands are going to disappear beneath the waves. And before they are enveloped by the ocean, these islands will become increasingly uninhabitable due to salination of the water supply, which makes agriculture less and less viable. It is likely that during this century the whole country will need to relocate permanently, affecting 500,000 people.

Around the rest of the world, rising sea levels will affect many more people. Even if whole counties don't disappear, many of their towns and cities will. The latest estimate is that 190 million people will be displaced by rising tides which cause coastal regions to erode, and whole towns and cities will be submerged by 2100. There will also be regular flooding and much more storm damage in the interior of countries. Huge storms that used to come once a decade will come once a year, and they will cause increasing damage to infrastructure. Once the storms come so often that the roads, electricity supply and water supply can't be repaired before the next storm arrives, then the places become uninhabitable. This will affect low-lying places like Florida and Mozambique first, and their populations will start to migrate. By some estimates they already are, not because the regular storm flood damage is unmanageable but because insurance companies are starting to refuse to insure properties.

So what can we do? Even if we were to stop emitting carbon dioxide now, the sea levels will continue to rise. This is because there is a lag between the cause and the effect. Now that we have increased carbon dioxide concentration above 400 parts per million, we have in effect turned the thermostat up on the planet. All those family trips in the car we did with my mum to museums, and the holidays when we flew to Greece, contributed to increasing carbon dioxide levels. All that invisible gas is still providing extra warming to the planet and will continue to do so until we get it out again. To get it down below 400 parts per million we need not just to stop emitting – we need to take the carbon dioxide out of the atmosphere. Unfortunately we currently have no viable plan for how to do this.

The people of the Maldives are not waiting to see if the

rest of the world comes up with a plan. The islands are on average only 1.5 metres above sea level and their highest point is five metres above sea level. They know what is coming, they understand the science, and they know that their islands will be submerged under the waves. Their plan is to build upwards, and they have already started. They are building an artificial island – called Hulhumalé – to give refuge to the population and protect them against sea level rises. The Maldivians started dredging sand from the bottom of the sea when I was hanging out in my yoga retreat in 1996. They have achieved a lot more than me in the intervening twenty-seven years – a whole new island has taken shape. It is one and a half miles long and half a mile wide.

Crucially Hulhumalé has been constructed five metres above sea level, with the capacity to house 240,000 people. The fabricated island has now been connected to the main Maldive island of Malé by an enormous 1.39-kilometre-long road bridge. Hulhumalé incorporates an international airport, and this is vital because the construction works and the bridge cost hundreds of millions of dollars to build and will need constant maintenance. Most of this was supplied by the Chinese, along with the construction expertise and equipment. The islanders plan to pay back this investment through tourism, the main industry of the nation. Of course, the irony is not lost on the island's inhabitants that in order to secure their future and not sink beneath the waves, they are relying on a form of transport that emits more carbon dioxide per mile travelled than any other. This is not their fault – it is currently their most viable option.

But is encouraging people to fly on holiday not going to make climate change a lot worse? My yoga instructor that day in Oxfordshire didn't seem to think so. We talked about it in bare feet while sipping green tea. I told them that every time someone asked me to clear my mind and go to my happy place I always ended up on a tropical island, but I was pretty sure that I hadn't arrived by boat, I had flown there. I had navigated a hot sweaty airport and emerged from a crowded terminal building, got into a taxi and so arrived at the slice of paradise. But was that right? Surely if I had to pollute the Earth's atmosphere by getting to my happy place, it shouldn't be my happy place? They told me I was over-thinking it, that I should relax my neurotic brain since no carbon dioxide emissions were created by imagining going to a tropical island. This re-assured me as I sat there sipping tea, watching the trees sway in the wind. But then my rational brain clicked back into gear as it had been trained to do as I am a scientist. But if people repeatedly dreamed of going to tropical para-dises, surely this would manifest in them actually wanting to go there for real. Wasn't this the basis of the whole travel industry and the constant marketing – to fulfil our dreams and desires of happiness by travelling the Earth and seeing all its wonders?

For most of history very few people travelled further than a few miles from their village or city. Travelling to other countries was something that was dangerous and difficult. For most people holidays were 'holy days' celebrated locally. The concept of travelling on holiday for ordinary people only became possible with the invention of the first world-changing gas technology – the steam railways. Then it became possible to safely travel with the whole family to far-away places, and to stay for a few days or a week and then return to work. The beach holiday was invented as a means of entertainment and relaxation. Travel by steam liner and then by jet aeroplanes opened the horizons even further, because these gas technologies made it possible at an affordable price. Now travelling to other countries to experience a different climate, a different way of life, a different culture or a tropical paradise is part of how we live. Most wealthy people have become like migrating birds, millions flying towards the sun during their annual holidays.

One of the many impressive things about the turbine jet engines that power modern aircraft is that they are very efficient. They turn kerosine, a form of fuel similar to diesel and petrol, into a jet of hot gas that pushes an aeroplane forward at speeds of typically 500 km/h. Their efficiency in doing this, along with the relatively low cost of kerosine (as we will see), means that journeys which would have cost thousands of pounds a few decades ago are now affordable. There is nothing out of the ordinary for a British person to escape the grey cold gloom of January by flying to a beach holiday in a tropical location such as the Maldives. This is no longer a trip of a lifetime, it is commonplace. It might even be described as a right of every citizen in rich countries. Certainly if such holidays were to be banned because they emit huge amounts of carbon dioxide into the atmosphere, there would be uproar. No government that implemented such a policy would ever be re-elected. Everyone would point out that global carbon dioxide emissions from the aviation sector are only 2.5 per cent of the total. There are much bigger problems to solve such as decarbonizing electricity generation and automobiles. Leave our beach holiday out of this, many people say. And so do the people of the Maldives, because without holiday makers flying thousands of miles for a taste of paradise, they can't fund the building work to rise above the rising sea levels. Without the tourist income they will have to default on their debts and perhaps even lose sovereignty of their islands.

So we are agreed then? The planes to beaches must continue to fly and beach holidays are here to stay. Only there is a snag. The numbers don't add up. When we calculate the carbon dioxide emissions per person due to all their activities, there is an inconsistency. For a person in the UK this is 9.66 tonnes of carbon dioxide per person. In the US it is 19.78, more than double that of a UK citizen. While for a person in China it is 4.58, half that of a UK citizen. For a person in Kenya, it is 0.3 tonnes per person. These numbers are calculated by adding up how much fuel a household uses for heating, cleaning and washing, travel, eating, clothes, TVs, electronics and other stuff.

When I plug in the numbers for me and my family, I find they come out at an average 6.97 tonnes per person – less than the UK average of 9.66 tonnes. But if I calculate just my own ... I come out at 11.72 tonnes. We share everything, and all our family holidays are by train because my wife refuses to fly. The extra 4.75 tonnes that I emit every year is because I fly to America and other places such as Asia as part of my academic work. This means I double my individual carbon emissions by flying. If everyone in the world flew as much as I do (four flights a year, 20 hours in the sky), the carbon emissions of the world would go through the roof. The calculation is easy to do: 8 billion people flying is 35 billion tonnes of CO_2. That's not 2.5 per cent of total global emissions – that's 10 per cent of global emissions. All that carbon will have to be removed from the atmosphere to stop the ice melting, the sea warming and global sea levels rising. So it turns out that flying to our happy places is not a big part of the problem if only rich countries do it, but if everyone does, then it's a big problem. And it doesn't seem fair to expect that as developing countries become richer, their citizens should fly less often than those of the rich countries. So what can we do?

The obvious thing to do is to stop flying until we create a form of aviation that has zero carbon dioxide emissions. Aeroplanes powered by electricity already exist. Small six-seater planes have been retrofitted with electric motors which rotate propellors and use batteries instead of kerosine. These typically have a range of 100 miles. Companies are now springing up to redesign planes to make them more aerodynamic and with lighter batteries, and this means that within 10–15 years all flights covering a range of 500–1,000 miles could be electric. If the electricity used to recharge the batteries comes from renewable sources, in the next few decades it is likely that short flights for small numbers of people will be affordable and sustainable.

But what about the Maldives? They are situated 700 miles

off the coast of India. At typical speeds of 200 mph such a journey would take three and a half hours. This seems just about doable for those living in the south of India, but for the rest of us it would require several hops over several days. The distance from London to the Maldives is 5,534 miles, so a trip to this happy place would require twenty-seven hours of travel, with a week of overnight stops to refuel. The cost of travel and the accommodation along the way would add to the price. In contrast, a modern jet plane flies direct to the Maldives from London and will get you and your family there in a little over eleven hours for £700 per person. Unfortunately the carbon dioxide emitted is also 2.52 tonnes per person.

The reason the price is this low is that modern jet aircraft are so energy efficient they can carry enough jet fuel to fly 400 people for eleven hours at 500 mph. For an electric aircraft to do that is currently impossible. The key difference is the amount of electric energy stored in a battery per weight. Jet fuel is a wonder material on this metric. It gives approximately 1,000 per cent more energy per weight than a battery. Unless there is a revolution in battery technology and aircraft design yielding a huge increase in energy efficiency, boarding an electric Dreamliner to take a non-stop flight to the Maldives is complete fantasy.

But what if instead we used a gas that is regularly touted as the fuel of the future: hydrogen? This gas can be liquefied and stored in the fuel tanks just like jet fuel. It can then be used to drive the jet engines by reacting with oxygen in the atmosphere to produce thrust for the aircraft. The only emissions would be water. The biggest technical challenge

The Airbus Zero hydrogen power concept passenger plane

is how to store liquid hydrogen safely on the plane. This requires the fuel tanks to be heavier than those used to store kerosine because hydrogen needs to be cooled to −252°C. These storage tanks need to be capable of keeping this cryogenically stored liquid insulated and pressurized for the whole duration of the flight. This presents a design issue because there is not enough room to store it with conventional aircraft design. Redesigning the whole chassis of the aircraft is the most promising option, and Airbus has already announced a design which it is working on called Airbus Zero.

There are huge engineering challenges ahead, but at least all the components are known technologies. Hydrogen has been used as fuel for space rockets for decades. By 2035, which is the earliest date that Airbus thinks it could be operating the Zero, it might be that other parts of the economy have also moved towards hydrogen as a fuel produced by renewable energy sources. For instance, hydrogen gas can be produced by using electricity from wind turbines and electrolysing water. This works because water is a molecule that contains two hydrogen atoms and one oxygen atom, with the formula H_2O. Passing electricity through the water releases the hydrogen gas and oxygen gas. Since water is plentiful on our planet and wind energy has zero emissions, this promises to be a way to produce sustainable aviation fuel. If such zero-carbon hydrogen (called green hydrogen) can be made cheaply and be part of a distribution network so that all major airports (including the Maldives) have a hydrogen supply, then Airbus Zero may rescue not just the aviation industry but also the people of the Maldives.

There are many unknowns here, and the question is what we should do while engineers and scientists get on with the task of decarbonizing the aviation industry by designing new electric- and hydrogen-powered aircraft. What is clear is that if we continue to emit large amounts of CO_2 into the atmosphere, it will increase the speed at which the Maldives (and all coastal communities) are going to sink under the waves. Yet we cannot abandon them either: they have to pay for the loans to protect their islands from the sea and their main industry is tourism.

Another way forward which still allows the people of the Maldives to survive is to create a carbon tax on aircraft emissions. This tax needs to be used to help all those around the world, including the people of the Maldives, to protect themselves against sea level rises and other effects of climate change. The tax would need to pay for ways to capture carbon dioxide from the atmosphere, ideally the same amount of CO_2 that is emitted by flying.

The simplest way to capture carbon dioxide from the atmosphere is by planting trees. Trees use sunlight to absorb carbon dioxide from the air and use it to grow. The Earth used to be covered in forests a few thousand years ago. Population growth used the trees as fuel for heating and cooking, as well as building materials for houses and sailing ships. In the process the forests were cut down. We moved on to use coal, oil and gas as fuels during the Industrial Revolution, but the trees were not allowed to grow back; instead the land was used to grow crops and maintain livestock. It is easy to say that existing forests should be maintained and more planted, funded by aircraft passengers who want to continue to fly, but the balance between the use of land for crops and forests for carbon capture creates tensions. Those who are poorer want the land to grow food so they can eat. Those who are richer want plentiful cheap food too, but they also want to go on holiday in aircraft. This competing tension is likely to play out this century as the seas continue to rise.

But even if land use issues can be resolved, is tree planting really going to solve the problem? There is a consensus that planting billions of trees is the cheapest way of capturing carbon dioxide. Analysis of land available for tree planting worldwide, which is land that does not already have crops growing on it, reveals that there is the possibility of planting 1.2 trillion trees. This has the potential of sucking huge amounts of CO_2 from the atmosphere, making a contribution to slowing global warming. There are two problems though. Firstly, the process of establishing the trees and for them to grow to maturity takes decades,

and the full effects of carbon capture would only be felt in
20–30 years' time. Studies that track these forests often
show they are cut down before maturity due to local politics or people not being paid to guard the forests. As we see
in places like Brazil, there are big pressures due to population growth that are leading to the desire to expand arable
land, which is leading to widespread burning of the
Amazon forests. Could it work that those who want to
continue to fly would be prepared to pay those local communities not to grow crops but instead make their living
by being custodians of the forests? We can expect a huge
growth in such financial schemes to tax the emitters of
carbon dioxide and the option to click 'off-set your carbon'
next to every airline ticket sold.

The second problem with planting trees to help the citizens of the Maldives facing rising sea levels is that even if
we could establish enormous forests this century to suck
up carbon, it is a one-off impact on the carbon in the atmosphere. After the forests are established and mature, then
these forests become carbon neutral – as many trees die
and become carbon dioxide as new trees grow to replace
them. The question we need to ask is how much would the
reforesting of the world reduce the carbon dioxide levels in
the atmosphere? Would it take the concentrations back
below 400 ppm? The current estimate is that planting a
trillion trees will reduce the amount of carbon in the atmosphere by 18 billion tonnes. This is just two years of current
carbon dioxide emissions. It's a complicated calculation
which takes into account how the soils, the oceans and the
forests all affect each other in the way they naturally absorb

carbon dioxide. But the upshot is that although planting trees is a good thing in terms of reducing carbon dioxide levels and creating habitats for wildlife, it will not bring concentrations back below 400 ppm.

Another card we can play to tackle rising sea levels is a technology called carbon capture. This is a technology that absorbs carbon dioxide emitted from power stations and converts it into a temporary form such as a liquid or a sponge-like solid. The material is sometimes then processed to scrub out the carbon dioxide and turn it into a stable carbon-based mineral a lot like chalk or limestone. The process requires energy and only makes sense if the energy used to remove the carbon dioxide from the exhaust is less that the energy produced by burning coal, oil or gas in the power station. Most of the current carbon capture technologies fail on this metric.

Even if we can get them to work, they don't solve the problem of aviation emissions, because carbon capture cannot be viably installed on an aircraft, not least because the plane would grow heavier and heavier as it collects the carbon dioxide coming out of the jet engine during flight. The alternative would be to locate such technology on the ground and make passengers pay for the equivalent amount they emit. In the case of a flight to the Maldives this would create a pile of approximately 2.4 tonnes per passenger – you can think of it as creating a lump the size of a car every time you fly. The process could be powered by renewable energy, but at the moment it is not viable because the costs are so large.

So what chance do the citizens of the Maldives have? At

The artificial island of Hulhumalé in the Maldives

present they are not great, I am afraid. At a minimum those who still choose to fly and to inflict huge environmental damage need to pay for remediation. At a conservative estimate this means an immediate tenfold increase in the cost of flying. Such a high carbon tax is needed because the scale of change needed is so great, and time available is short. Part of this carbon tax would go to the citizens of the Maldives to help them build their island ever higher. They will need to do that because the levels of carbon dioxide in the atmosphere will still be way above 400 ppm and so the sea levels will continue to rise due to all the emissions from cars, food production, homes and industry – until these things all become net zero emitters.

To actually stop the ice melting and the sea levels continuing to rise, we need to suck all carbon dioxide that we have emitted over the last hundred years out of the air and put it back in the ground. Planting a trillion trees and paying local communities to be custodians of these forests, and the biodiversity they safeguard, will be important but nowhere near enough. We will have to develop carbon capture technology that is cheap and efficient. For the Maldives and other low-lying countries to survive, we are going to need to invent one soon.

It's an ambitious task but I think we can do it. It is the politics of those with vested interests in fossil fuels that are holding us back. Understandably, just like the tobacco industry before them, they are addicted to the money their industry makes from its alluring and yet destructive products. Some might say they are possessed by evil spirits. But as the seas continue to rise, and storms, heatwaves and droughts get ever worse, reason will eventually triumph over greed. This will take the form of high carbon taxes on flights, and other carbon-intensive activities, with money going to accelerate decarbonization and carbon capture, and to compensate those worst affected by climate change. The reason for my optimism is that we have been here before with another atmospheric gas, nitrogen. There is much we can learn about grass root change from that dark episode in our history, as we will see next.

12. Ghostly

Sometimes in the middle of the night when I can't sleep, I
go to the kitchen to pour myself a bowl of sugar-coated
nutty breakfast cereal. I then pad down the hall to watch
TV. If I'm lucky an old war film will be showing on some
obscure TV channel, and I can lose myself in it while I
munch and slurp a couple of helpings of cereal. As a kid
these films would excite me and my brothers beyond measure. They hit some deep spot inside us. Once the film had
ended it was as if we were possessed and had to run outside to play war. We would imitate the sounds of different

types of machine guns while spraying each other with imaginary bullets. I died many times at the hands of my brothers' superior fire power, collapsing onto my mum's flower beds and letting out blood-curdling screams. This annoyed my mum a lot. But it never occurred to any of us that this might be distressing for my dad, a refugee from the Second World War, who had so much of his family actually wiped out by war.

The Nazis tried to destroy my family because we were Jewish. Only a few of my relatives survived, including my grandfather and grandmother, who somehow escaped from Germany to Britain in 1939 and then managed to get my father, aged eight, to Britain too. It was a miracle really, given how many millions of other Jewish people were exterminated by the Nazis. So someone with my background really shouldn't love war films. I shouldn't love the sight of tanks rolling into a city while brave defenders take cover in a destroyed building. I shouldn't be delighted by the sight of a hero running towards the enemy's trench and lobbing a grenade into it. I really shouldn't happily munch breakfast cereal in the middle of the night while a soldier has his throat cut. And yet I do. The story of my breakfast cereal is also connected to war, by the way, and my love of it exposes other inconsistencies. It is the story of nitrogen and gas warfare.

By 22 April 1915 the First World War had been raging for almost a year. Trench warfare between Germany and the Allies was ongoing: it was a brutal combination of artillery bombardment and the regular massacre of troops as they were sent out to face machine gun fire from the

The first use of chlorine gas in the First World War

enemy trenches. The war was at a stalemate, with both sides gaining no ground and slaughtering each other's soldiers. On this particular day, however, there was something new, and lethal: chlorine gas. A scientist, Fritz Haber, an expert in chemical engineering, had solved the many technical problems with isolating and concentrating chlorine gas into canisters. These canisters were now deployed on the front line.

It is easy to detect chlorine gas in the air: it has a yellow/green colour and the recognizably sharp smell of bleach. One breath of it and you will start choking, a few more breaths and it burns your throat and lungs. Further exposure delivers an agonizing death. It kills anyone without a gas mask, whether a soldier or not, and this is one of the reasons why gas warfare had been outlawed prior to the First World War by the Hague Convention of 1907. But the stalemate and horrific losses on both sides had persuaded the German high command that the use of this weapon would bring victory so quickly that it would actually save lives – a recurring theme in warfare, as this reasoning led to the use of atomic weapons by the USA against Japan in the Second World War. Well, Fritz Haber was of this opinion too. He thought that there was little moral difference between killing an opponent with a gas molecule rather than a slug of metal from a machine gun. So on 22 April 1915, with the wind blowing in the right direction, the German troops released Fritz Haber's 168 tonnes of chlorine gas from steel cylinders along the front line at Ypres. Thousands of soldiers were killed and injured by the chlorine. They had no defence against it until gas masks were manufactured and supplied by which time both sides were using poison gas weapons. Wilfred Owen, a British soldier who experienced gas warfare, wrote a poem about it called 'Dulce et Decorum Est'. The second verse captures the horror of this type of warfare:

> Gas! GAS! Quick, boys! – An ecstasy of fumbling
> Fitting the clumsy helmets just in time,
> But someone still was yelling out and stumbling
> And flound'ring like a man in fire or lime. –
> Dim through the misty panes and thick green light,
> As under a green sea, I saw him drowning.

Germany eventually lost the First World War in 1918 and had to make reparations to the countries that suffered. But Fritz Haber emerged a hero in Germany, despite being dubbed 'the father of chemical warfare'. He was even awarded the Nobel Prize for Chemistry. This was not in recognition of developing a technique to kill people using chlorine gas, but for discovering a process to solve world food shortages, a process that all scientists of the day thought impossible. It is a process that would end up making breakfast cereal very cheap.

Food contains essential nutrients that we need to survive. Our cells and the cells of all living creatures are built using molecules called proteins. They are complex combinations of amino acids that are composed of carbon, hydrogen, oxygen and nitrogen. Our DNA is made from these proteins, and we use them to constantly build and repair ourselves. We get the ingredients to build proteins from the food we eat, from the cells of fruit, vegetables, cereals, nuts, fish, meat and dairy products. Which is fine, except that there is only so much to go round, which means that a typical ecosystem can only support a certain number of organisms with enough nutrients. In particular, nitrogen tends to be in short supply, and without it proteins cannot be built, and therefore growth stops and life is stunted. Other minerals are important too, such as phosphorus and potassium, but all life needs nitrogen. It's a self-limiting factor that restricts the populations of all living things.

The Britain to which my family fled as refugees was a land of plenty and still is. This is because it is blessed with large areas of land with fertile soil. This soil was established by the vast forests that covered the land before humans arrived. The trees in those forests harvested energy from the sun to build their branches and produce the oxygen we breathe as a by-product. The trees obtained the nitrogen they needed from the soil and returned it to the soil when they died. This cycle nurtured the other organisms in their ecosystem too: worms, flies, birds, squirrels, deer and humans all got their nitrogen from the same source. Either directly by eating plants or indirectly by eating each other.

The need to get nitrogen from the soil is odd in a way, because we are bathed in nitrogen – the air is 78 per cent nitrogen. But this is atmospheric nitrogen and can't be extracted from the air we breathe. This is because the nitrogen gas is in the form of a molecule in which two nitrogen atoms are bound together, denoted $N_2(g)$. The bond holding the two nitrogen atoms together is a triple bond in which each atom shares three electrons. This bond, which has been called 'the strongest bond in nature', is very hard to break. For this reason the nitrogen gas in air doesn't react or interact with pretty much anything else: all its available electrons are bound up inside this mighty chemical bond. Thus although there is a lot of nitrogen on the planet, most of it is chemically unavailable to us, and so the stuff that does exist inside soils in the form of nitrates and amino acids needs to be preserved and recycled.

It is the soil which does this recycling, and in so doing nurtures us all. The soil contains a vast range of bacteria and fungi, as well as minerals and animals such as worms and insects. These organisms live in the soil and consume the dead plant matter, gaining energy from it and the other dead creatures. This sounds grisly, but death is a vital part of life on this planet – we can only live through the death of other organisms. The recycling of dead matter makes the soil a medium rich in nutrients, such as nitrogen, which allows new plants and trees to grow and so keep a forest healthy.

When humans cut down the forests in Britain and other parts of the world to move from a hunter-gatherer lifestyle to an agricultural one, they radically changed the ecosystem and had to learn how to manage the fertility of the soil. By eating the crops that grew from the soil they were extracting nitrogen. For this to be sustainable, they had to find a way to return the nitrogen to the soil; otherwise its fertility would decrease, crops would fail and famines would ensue.

Early farmers tried to solve this problem by performing religious rites which often involved sacrificing animals to appease the fertility gods. If the crops failed and the community starved, it was thought to be some punishment from the gods, and more sacrifices were needed. Equally when the crops were abundant it was a clear sign that the gods were looking kindly on them. Worldwide pretty much all early civilizations prayed to fertility gods as a way to come to terms with the invisible cause and effect, such as the action of nitrogen, that determines soil fertility. Over time though, farming communities realized that there were more effective approaches to managing soil fertility than praying. The most important was recycling animal dung and mixing it with crop stalks and other plant residues containing nitrogen and other nutrients. This process is called composting and it is still how organic farmers nurture and maintain fertility of the soil today. It is also how gardeners maintain the productivity of their soils.

The process of making compost is not difficult: nature

does most of the work for you. I live in a flat in Central London with a roof garden, and I have been making compost on our roof for over twenty years. None of our food waste that my family produce goes into the dustbin. Instead we collect this nutrient-rich material in a small bucket. Every couple of days we transfer this bucket into one of two large containers on the roof garden. Inside these composters is an ecosystem of earthworms, bacteria, fungi and thousands of other organisms that live off the constant supply of nutrients we provide. Over a few months they consume our food waste and recycle it into a nutrient-rich soil called compost. It never ceases to amaze me that this process works so easily. Nature needs no instruction, no equipment, it just gets on with the job that it has evolved to do. Into the composter go old bits of carrot and half-eaten bread rolls, and out comes lovely dark earth. It is a piece of magic worthy of a fertility god. In our family we refer to our food waste as offerings to 'the god of compost', and you'd be right to detect the influence of Mum in this. Thanks to her, not offending the god of compost is part of my DNA. When away from home I can't bring myself to throw away food waste, and often bring it home so we can compost it and use the soil produced to grow flowers, tomatoes, beans, herbs and trees. Which is what we do on our roof.

Just to reassure you in case you are worried for my neighbours: the compost process doesn't smell bad. It smells, well, earthy. The strongest smells wafting around our flats are usually the smells of people frying fish or smoking cigarettes on the roof.

But however clever and satisfying composting is, it still

doesn't solve the problem that there is only a certain amount of nitrogen and other nutrients to go around. To support a growing population more food needs to be produced from the same amount of land. This problem has a biological solution too.

A few bacteria have evolved ingenious chemical pathways to crack nitrogen's triple bond and create soluble nitrogen compounds from nitrogen gas. The most important of these are rhizobium bacteria. They turn atmospheric nitrogen into ammonia, which is a molecule that plants can then use to create proteins which are stable and fix nitrogen into the ecosystem for other organisms to use. Some plants have evolved a symbiotic relationship with rhizobium bacteria, creating nodules in their root systems where they deliver nutrients to the bacteria in return for the nitrogen. Peas, beans, clover and other legumes do this. When you plant peas, they recruit rhizobium bacteria to their root systems and fix nitrogen into the soil, making it more fertile. Farmers knew the importance of these crops as far back as 8,000 years ago. Staple crops such as wheat, rice and potatoes don't do this, but they can still benefit from the extra soil fertility if they are planted in soil where clover, peas and beans were grown the year before. Thus rotating the planting from year to year while adding compost maintains soil fertility. This crop rotation is a clever system which also prevents certain pests and diseases establishing themselves in the soil. It does so by removing the particular food source (perhaps wheat) from these pests for that year, and thereby decreases their prevalence in that field. Instead a different crop, such as clover, is planted in that field, which adds to soil fertility by recruiting rhizobium bacteria to its root systems and increasing nitrogen. Meanwhile the wheat is planted in a different field. A typical crop rotation in the UK during the years when it was perfected in the 1770s was a four-field crop rotation of wheat, turnips, barley and clover.

The processes of crop rotation and the application of compost onto the land are human technology. It is not natural, but it does maintain soil fertility. However, it seems that soil can never be fertile enough to satisfy humans. This is because, historically, as communities become more prosperous, their food demands always increase. This problem was elaborated on by Thomas Malthus in *An Essay on the Principle of Population*, published in 1798 in England. In this book he argued that improvements in farming techniques, such as crop rotation, never really help the poor because the greater abundance of food causes the population to increase. This then has the effect of reducing the amount of food per person, which creates more poverty. Malthus argued that this was inevitable, and population pressures would always create conditions for famine, more plagues and even war.

My habit of helping myself to cereal while watching late-night TV would have amazed Malthus. It is not just the abundance of cereal but the fact that most of the UK population can afford to eat as much as we like. The UK is self-sufficient in wheat and other cereal grains, despite having a population of almost 70 million. This is nearly a tenfold increase in the population since Malthus published his book. In fact the biggest health problems for most citizens in the UK come from eating too many processed foods such as breakfast cereal. So how did this happen? How did we avoid the Malthusian trap? The answer is our ability as a civilization to obtain nitrogen gas from the air in a form that plants can digest – this process creates synthetic fertilizers.

After Malthus various experiments were conducted to artificially increase crop yields, and once nitrogen was identified as a key ingredient, the world's geologists started looking for deposits of nitrogen-rich minerals. They found them in the Chincha islands off the coast of Peru, which had been home to enormous numbers of nesting sea birds which pooed for so long on these islands that this 'guano' was knee deep and extremely rich in nitrogen and other nutrients, like phosphorus. It was not just the birds that were responsible for the guano: Huamancantac, the Inca god of guano, presided over its production to help the Inca civilization fertilize its crops and grow food.

Once colonialists from Europe and America found out about this, they used sailing ships to import guano from halfway across the world to fertilize crops at home. Between 1830 and 1900 the total population of these countries expanded from 300 million to 500 million. The consumption of grain per person also increased, as the populations grew richer and fatter. Significantly for me and those who eat cereal at night or at any other time, this was when (1894) John Harvey Kellogg and his brother, W. K. Kellogg, invented the first breakfast cereal: Corn Flakes.

In the 1850s the UK was importing 200,000 tonnes of guano a year, and wheat yield per hectare doubled. Such was the mania around this source of nitrogen that there was intense competition to dominate the trade. Spain even went to war with Peru over who owned the Chincha islands. Of course, the guano, a limited resource, was bound to run out, and so it did. Further nitrogen supplies were sought and discovered; they were rare and never very plentiful. The reason for the rarity was clear: to be useful as a fertilizer, the nitrogen compounds in the deposits have to be soluble in water, to allow plants to take them up through their roots. But when 70 per cent of a world is covered by water, with plentiful rain and rivers, finding places where nitrogen-rich minerals have been accumulated through the build-up of bird poo, but have not dissolved away, is always going to be difficult.

A mineral deposit of caliche, a sodium-nitrate-rich rock, was discovered between Chile, Bolivia and Peru in the extremely arid plateau of Tarapaca (it rains once in 3–10 years). The exports of nitrates to Europe and North America accelerated between 1830 and 1880, and were valuable enough for a Nitrate Steam Railway to be built to carry the mineral to ports on the Pacific coast. The demand was so great that in 1874 Chile went to war with Bolivia and Peru to establish control over the exports. It took five years, but eventually Chile won control of the desert and so won one of the world's only sources of nitrates, yielding millions of dollars in taxes for the Government of Chile. The nitrogen was not only increasingly demanded by farmers, but also by those making explosives. It is the key ingredient in gunpowder and nitroglycerine, where the nitrates react to create nitrogen gas in an extremely violent reaction (this is also why stores of nitrogen fertilizer occasionally blow up).

Towards the end of the nineteenth century, there was a realization that the major powers of the world were beginning to rely heavily on a single major source of nitrates: Chile. Access to this source increasingly determined their ability to feed their dense populations and to wage war or defend themselves. The leaders of Germany were particularly anxious about this. At the beginning of the twentieth century, it was the world's largest importer of Chilean nitrates, but due to its geography and the location of its sea ports, it was susceptible to naval blockades which would deprive it of access to nitrates. This is the background to the life of Fritz Haber, the father of chemical warfare, who in the early twentieth century was dedicating all his efforts to solving the problem of harvesting nitrogen from the air.

Fritz Haber was born Jewish and went to a mixed religious school open to Catholic, Protestant and Jewish kids. He didn't identify as being Jewish but saw himself as a German patriot. He was determined to prove himself to his father who was the founder of a chemical-engineering business making dyes, paints and pharmaceuticals. As he developed his chemistry career, he clashed with his dad, and then later with the scientific establishment, in particular with Walter Nerst, who was one of the great chemists of his generation. Nerst analysed some of Haber's early promising results and publicly ridiculed them, implying that Haber and his team were shoddy scientists pursuing a pipe dream of capturing nitrogen from the air. In fairness to Nerst, it was not just vindictiveness, the majority of chemists agreed: it didn't seem remotely possible to break the mighty nitrogen triple bond to create fertilizers.

Haber, though, ignored them all, taking the criticism as fuel for his obsession. His idea was to use explosive hydrogen gas as a way to get nitrogen gas out of the air. The chemical reaction between hydrogen and nitrogen produces a strong-smelling substance called ammonia. It has the instantly recognizable smell of urine, which even in small concentrations wafts up your nose and seems to slap your olfactory receptors. Urine's ammonia content makes it a nitrogen fertilizer, and urine has been collected for this purpose since ancient times. People peed onto their compost to enhance its nitrogen content. The Romans traded in urine, calling it liquid gold. They used it for many chemical processes, including making cleaning products such as toothpaste (modern bleach is also an ammonia product). The Chinese invented gunpowder using urine as a source of ammonia to give it its explosiveness. So Haber knew that if he could make ammonia from the nitrogen gas in the air, it would be an invention that would change the world.

The chemical reaction between hydrogen and nitrogen gases is very inefficient though, and Haber had to use high temperatures and pressures to make it work. Nevertheless, Haber was convinced he was onto something. Why? Well, ammonia gas dissolves in water. So the small amounts of ammonia created could be dissolved. This ammonia solution would then become a reservoir for the nitrogen. Problem solved. Or at least in theory. Would it be cost effective given the high temperatures and pressures needed?

Haber's experiments showed that 0.0125 per cent ammonia could be produced, a tiny amount but still promising enough to justify the large energy expenditure of heating up the gases, putting them under pressure. Haber declared that it could work thermodynamically and economically. Nerst looked at these experimental results, and pronounced them to be laughably wrong. Nerst went further, publicly humiliating Haber, calling his experiment 'strongly inaccurate'. Haber and his team repeated their experiments and found Nerst was annoyingly right. Haber became ill and almost abandoned the whole enterprise.

But then Haber realized that the reaction between nitrogen and hydrogen could be made to occur faster on surfaces of metals called catalysts, which speed up chemical reactions. This is how the catalytic converters work on a car exhaust: small particles of platinum and rhodium are embedded into the exhaust pipe and accelerate the reaction of nitrogen oxide gas into less harmful products before it leaves the exhaust pipe to pollute the air. Fritz Haber thought he could find a metal catalyst to accelerate the reaction to create ammonia.

Another reason Fritz Haber continued with the research was because he was a proud German. He knew that a future of prosperity and high crop yields to feed a growing population needed a secure and cheap source of nitrogen fertilizer. He also knew that ammunition and fire power for its armies depended on plentiful and cheap nitrates. He wanted to be the person to provide that for his beloved Germany.

Fritz Haber made it through the humiliation of the early failure, and with the help of his research team showed that

An example of a fertilizer production plant

insanely high pressures and high temperatures (600°C) yielded 6 per cent ammonia, a hundred times more than his earlier experiments. The catalysts he used were based on rare metals such as osmium and uranium, and increased yields to 10 per cent. By 1909 Haber was ready to commercialize the process with the German company Badische Anilin und Soda Fabrik (BASF). He and his team demonstrated the lab process to BASF's management, showing that the only inputs to the process were the gases nitrogen and hydrogen, while ammonia came out the other end, steadily and continuously hour after hour. For the first time in human history, a constant source of soil fertility was possible out of thin air.

BASF, under the talented direction of the engineer Carl Bosch, designed and built the industrial process, which dominates the production of synthetic fertilizer worldwide and is still called the Haber–Bosch process. It is very energy intensive due to the high temperatures and pressures of the gases involved, which is why the early chemical plants were sited near coal supplies. By 1913 the massive ammonia plant at Oppau was operational and producing 30,000 tonnes of ammonia per year (enough to fertilize crops to feed 5 million people). This seemed a human triumph, fertilizer need never be in short supply again. Birds could poo in peace, knowing that their crap didn't need to be collected and shipped halfway round the world. High crop yields could be ensured and no one would ever starve again. Hurray!

But it was a triumph accompanied by a spectre that has followed the Haber–Bosch process ever since. In 1914 the First World War broke out, and the immediate stalemate on the Western Front initiated trench warfare and demanded an enormous amount of munitions. A naval blockade deprived Germany of Chilean nitrates with which to make explosives. BASF stepped in to help by turning the Haber–Bosch process into a production route for explosives, keeping Germany in the war until 1918.

Fritz Haber was named as a war criminal. When his ammonia synthesis work won him the Nobel Prize for Chemistry in 1918, this understandably caused an international outcry. But Haber accepted the prize, defending his inconsistency by maintaining that his chemical-weapons work was carried out to save his country from extinction.

If the ghosts of all the 100,000 soldiers who died as a result of gas warfare in the First World War had crowded into that Nobel Prize ceremony as Fritz Haber was honoured, first among them would have been the ghost of his first wife, Clara Immerwahr. She was a brilliant chemist in her own right but committed suicide when she heard that her husband had been the one to champion the use of chemical weapons. Little did she know, though, that his synthetic fertilizer was also a kind of chemical weapon that would wreak a different type of destruction on the world.

The Haber–Bosch process is now the dominant form of nitrogen fixation for agriculture. All countries use it. Most farmers use it. It has made agriculture worldwide far more productive, and food cheaper. We all rely on it. We really would not have plentiful food without it. The breakfast cereals I eat late at night would be an expensive treat without nitrogen fertilizers. By the last estimate there would only be food available for half of the population of the world without the Haber–Bosch process. You would have thought that inventing a process to obtain the most important nutrient from thin air would enrich life on Earth, but unfortunately it has side-effects that were not anticipated.

The use of synthetic fertilizer has changed agriculture radically. Huge fields of wheat, corn and rice are planted every year without the use of crop rotation or of compost. To protect those vast crops from weeds and insects, pesticides and herbicides are used to kill off anything else on the field. This has turned the soil into an industrial landscape optimized for single-crop production. Other slow-growing and low-nutrient species of plants have been systematically eradicated. The fields have become a desert for organisms such as worms and micro-organisms in the soil that would normally regenerate the fertility. The loss of these creatures in the soil leads to compaction and has a knock-on effect on other ecosystems, such as bird and insect populations. The populations of these have been decreasing radically since intensive agriculture began in the twentieth century. When I was a kid, my parents would take me and my brothers camping in the summer. Without exception every trip we took resulted in the windscreen of our car being covered with splatted insects. This is because there were so many of them, and just driving down the road caused us to collide with them. Now a generation later, when we take our kids camping, there are no insect splats on our car's windscreen. None. The vast decrease in insects has caused the vast decrease in birds, animals and many other parts of our ecosystem.

The nitrogen injected into fields that have little life other than a single crop of wheat destined to become breakfast cereal has another effect on the environment. The nitrogen is soluble in water and when it rains these nitrates run into streams and lakes. This fertilizes the algae in the water and causes them to grow. In growing, these algal blooms remove oxygen from the water and thus kill aquatic life. The net result is huge numbers of dead fish and other aquatic organisms. The problem has been getting more serious every year since artificial fertilizers were invented.

It is not the use of fertilizers that is inherently bad: nitrogen fertilizers are an important part of our life support system. We would all be paying a lot more for our food, and half of the world would experience food poverty, if nitrogen fertilizers did not exist. But the price of cheap and plentiful food has undoubtedly been a loss of the Earth's biodiversity. If we don't transform farming into an enterprise that safeguards soil health and biodiversity, then we will have failed future generations. Our kids will not live in green and pleasant lands but increasingly in deserts, haunted by the ghosts of people like Fritz Haber who unwittingly unleashed this chemical weapon into our farms.

Things are changing though. Those in rich countries are starting to voluntarily pay more for their food in order to prioritize biodiversity and protect the climate. Paying extra is necessary because at the moment it is hard for organic farmers, who do not use artificial fertilizers or pesticides, to compete with intensive farming. The yields per acre are lower because available nitrogen and other nutrients are lower. They are lower because the crops feed the organisms in the soil as well as insect populations, birds and other animals. By buying organic food people are paying to support the biodiversity of the land. Governments are also realizing the value of paying farmers to be custodians of ecosystems as they realize that if they don't incentivize the preservation of the natural environment, the impact of climate change on crop yields will be much greater.

Many different solutions are offered to solve the problem of feeding the human population while maintaining biodiversity, but they all in some way involve us as a species becoming careful in our ability to manage nitrogen and other nutrients. That means understanding how new agricultural practices interact with nature through the invisible cause and effect nitrogen has on our ecosystems. We are living through the Great Nitrogen Event, where nitrogen is being made available to life forms at an unprecedented scale. Similar to the Great Oxygen Event billions of years ago that gave rise to us oxygen breathers, it will change the nature of life on the planet. Things will never be the same again, but if we want to keep control of the nitrogen cycle, it will need to become part of how we organize ourselves on this planet. Like controlling the climate, it will make gods of us, because if we succeed, we will have planetary power. Are we ready for that?

The Hereafter

We evolved on a planet seemingly inhabited by ghosts, will-o'-the-wisps, fertility gods, sprites, fairies and many other strange and mysterious entities. For thousands of years they dominated our religions, culture and folklore. They also seeped into our subconscious. With the development of science and in particular the discovery of gases, we realized that the actions of these invisible supernatural entities had rational explanations. By doing so we largely banished wind gods, phantoms and sprites from the realm of normality. Gas technologies such as engines, bicycles, vacuum cleaners, automobiles, aeroplanes, fertilizers, anaesthetics have replaced them as astounding influences in our lives, creating a new type of supernatural world, one we have invented, and one in which we wield god-like powers due to our engineering prowess.

Although we now rationally understand gases and their influences on us, we are emotionally attached to the age-old myths that surround them. We are creatures for whom the invisibility of cause and effect spurs speculation about supernatural forces or conspiracy theories. So what does the future hold for our relationship with these invisible substances, and the ghosts and gods that accompany them?

Personally, I am thankful for the ghost of my dear mum. I detected her perfume on buses and caught sight of her in a crowd several times. Returning to the place I planted a tree for her after she died, I felt her admonishing me for the mud on my trainers (it's a field, Mum!). Most significantly she made her way into the pages of this book without my conscious bidding. Her appearance transported me back into childhood and my formative years, when I both absorbed and rebelled against her philosophy of life. Many would say that what I am describing is just memory and emotional associations of her, perfectly normal, and not the action of an actual ghost. They might say I have an overactive imagination. Perhaps, but is not 'ghost' a better word, and perhaps a more accurate one, for the real influence that the spirits of the dead have on us?

The word 'ghost' has power precisely because although

we have discovered the rational truth behind the invisible forces in our lives, it has also made the invisible more real. Mobile phone signals really do exist, they really are powerful, and they really do travel through our bodies and homes bringing information, data and live video to us. It is amazing that the invisible world can do all this, and so the irrational part of our minds takes note. *What else could be going on?* it speculates. My prediction is that the future will be as full of ghosts and gods as the past; they represent one way in which our irrational minds constantly seek to resolve the mysteries of the world, sometimes in opposition, but often in addition to, rationality.

Talk to modern sailors and it won't be long before you find that many of them are superstitious about going to sea. It is a risky business. The seas and oceans are driven by capricious winds and storms that can only be predicted a few days ahead, and, even then, with a low degree of accuracy. Rationally the survival of a boat in a storm depends on numerous factors. Each one can be analysed, but nevertheless it is easier for survivors to say they were lucky and give thanks to gods or a lucky charm. These superstitious beliefs are a recognition that for all our knowledge about the wind and the oceans and how they work, our control over them is tiny. This irrationality is not a human defect, it is an evolved capacity to help us deal with the insecurity of living in an unpredictable world. Rituals involving gods or charms are a recognition of this, they make emotional sense. Similar behaviour is mirrored in many aspects of our lives, including of course, in my case, the visitations of my dead mum.

The acceptance of our irrational minds hasn't stopped us trying to rationally understand gases and use them to have superpowers. We flick a switch and the kettle boils water almost instantly due to the currents of electricity flowing into our homes, all of which relies on gas technologies described in this book. Our homes are made toasty warm by another gas technology, methane. More fundamentally the steam engine catapulted us into a world of automated manufacturing. This made goods cheap and plentiful, allowing societies to become materially wealthy. The fact that billions of people can afford plentiful clothes, phones, TVs and cars is due to a century of manufacturing and cheap energy in the form of fossil fuels. The gas technologies of agricultural machinery and synthetic fertilizer freed the majority of the population from having to work the land manually and gave us a world of plentiful food.

Gas technologies in the form of sailing boats, steam trains, automobiles and bicycles opened our horizons. Dreams of flying became reality through the development of hydrogen balloons and airships. Then came aircraft and the development of the jet engine, which gave us the superpower to fly to any place on the planet. These gas technologies were first wondrous and expensive, and then, miraculously, became part of everyday life. They have given us wealth, freedom and liberation. Like birds we have become migratory beings – for us, getting into a car, on a train or on a plane is more normal than staying in one place. It is because of gas technologies that we have ceased to be people who live, work and die in the same village in which we were born – we are now global citizens.

We rely on gases to keep us healthy. No modern hospital is complete without tanks of oxygen. It saves millions of lives a year, as epitomized by the COVID-19 pandemic. Laughing gas alleviates pain during childbirth, dentistry and other medical procedures. Argon laser treatment restores eyesight miraculously and painlessly. Vacuum technology is at the heart of x-ray machines and electron microscopes, while a continuous supply of helium is essential to run the MRI machines that magically look inside us to reveal our ailments and diagnose cancers.

Walking out of a hospital one day you might take a deep breath and experience one of those moments of profound appreciation of being alive. That deep breath is of course thanks to the presence of air in our atmosphere, the ultimate life support system of the planet. In that breath is the oxygen we need to live, the nitrogen that provides soil fertility, the carbon dioxide that keeps the planet at the right temperature for living organisms, and the winds that bring fresh water from the ocean. Their concentrations have varied widely over billions of years, mirroring the rise and fall of different forms of life and their influence on the climate. At first there was lots of carbon dioxide and no oxygen. It was only the emergence of photosynthesis, which used up the carbon dioxide and produced oxygen as a waste product, that changed this. Since then organisms producing oxygen and those producing carbon dioxide have been in a global dance, each relying on the other for their fuel. When that balance tips, we know from the fossil record that dramatic things happen, such as large changes in global temperatures and mass extinctions. For instance,

500 million years ago carbon dioxide was at concentrations of 5,000 ppm; it is now at a concentration of 420 ppm. Where has all this carbon gone? It is in the Earth's crust in the form of carbonate rocks, coal, oil and methane gas. This reduction of carbon dioxide in the atmosphere reduced global temperatures by 10°C and made it habitable for us. Currently we are reversing that process by burning the coal, oil and methane. Finding an economical way to stop doing this and put that carbon back in the ground is the biggest technological and political challenge humans have ever faced. I am confident we will do it. How hot it gets before we achieve this depends on how fast we can persuade the fossil fuel industry that it will be destroyed too if it doesn't stop. Even though much of what needs to happen to reverse climate change will be invisible to us, we can keep an eye on progress very easily. When politicians and the fossil fuel industry say they are making progress, ask them what concentration of carbon dioxide is in the atmosphere. At the moment it is 420 ppm. If it goes up, our life support system is in trouble; if it goes down, we are making progress.

We also need to get the nitrogen cycle under control. The Great Nitrogen Event, which we initiated through the global use of synthetic fertilizers, has given us enormous crop yields, allowing us to feed the growing global population. But it has also disturbed the equilibrium of the amount of nitrogen-based nutrients available in the environment. Their use in intensive farming is destroying biodiversity at an unprecedented scale. Nitrogen pollution is ending up everywhere, in rivers, lakes and oceans. It causes algal blooms which use up oxygen and water which then cause the death of vast numbers of fish and other organisms. The nitrogen is also released into the air in the form of nitrogen oxide gases. These gases are potent greenhouse gases and significantly contribute to climate change. They travel long distances in the wind and then cause acidification and nutrient pollution of soils and ecosystems, seriously affecting bird and insect populations. Studies also show that this nitrogen pollution is being linked to cancers in humans. So as with carbon dioxide, although nitrogen is part of our life support system, we need to get it back in balance if we are to thrive on this planet.

We cannot survive without air, but every breath we take not only brings oxygen but also tiny particles, harmful gases and tiny droplets of water sometimes containing bacteria and viruses. The purity of the air we breathe, especially in cities, has always been a contentious topic. At first it was full of foul smells from human and animal excrement mixed with wood smoke that choked inhabitants. This miasma was blamed for all sorts of illnesses and the spread of diseases. In the end we discovered that some terrible diseases are indeed transmitted through the air, such as TB and COVID. Smoke causes lung damage and brain damage, and makes asthma sufferers' lives a misery. Through the development of cleaner forms of energy such as electricity and gas, wood and coal smoke has been eliminated from many cities. This success was short-lived though, since smoke pollution was replaced by another invisible killer: the small particles in the exhaust from cars and tyre particles which become airborne. Cities such as London are now bringing in ultra-low-emission zones to deal with this issue, with taxes on highly polluting vehicles and speed limits which reduce tyre wear. However, studies show that an increasing amount of the pollution that envelops towns and cities is nitrogen oxide gases and ammonia being blown in the wind as a result of intensive farming.

This issue of sharing our atmosphere with all people on the planet and with all other life forms is perhaps our biggest challenge as a species. It is our common life support system and yet we have only fragile political and economic tools to help us cooperate to preserve it. Focusing solely on carbon dioxide is foolish, as I hope is clear from this book: it is part of an intricate web of invisible interactions with the other gases and living organisms. We are bound to cause further unexpected side-effects if we treat carbon dioxide as an isolated problem to be monetized and traded and dumped into the seas, on land or underground. The story of fertilizer and pulling nitrogen gas from the air on a global scale shows us just how badly wrong it can go. Current farming practice has a significant impact on global warming through its use of fossil fuels to make fertilizer to create ever more intensive farming methods. In return, climate change reduces agricultural output through droughts and floods, which creates a demand for more fertilizer in a dangerous feedback loop that we are seeing playing out now.

Thankfully we know what to do to reverse this problem. There is a global movement to shift away from intensive agriculture and replace it with regenerative farming. Governments and industries have a huge part to play but know that everyone who has a choice about what they eat can get involved. It requires you to buy food from farmers who are managing their land to absorb carbon dioxide, control nitrogen and support biodiversity. The food produced is no less delicious than food made by intensive farming, but you will have to pay more for it. These are often labelled 'organic', 'free range' or 'regenerative'. What you are paying for is to preserve something invisible – the life support system of the planet. That of course makes it a tricky sell. But having read this book, hopefully you will be receptive. Similarly, choosing low-carbon forms of travel won't make you a less interesting person, and future generations will thank you for it.

Further Reading

S. L. Bridle, *Food and Climate Change without the Hot Air*, UIT Cambridge Ltd, 2020

Steven Connor, *The Matter of Air: Science and Art of the Ethereal*, Reaktion Books, 2010

Thomas Crump, *A Brief History of the Age of Steam: From the First Engine to the Boats and Railways*, Avalon Publishing Group, 2007

Arthur Conan Doyle, *The Adventures of Sherlock Holmes*, Penguin Classics, 2018

David E. Fisher, *Much Ado about (Practically) Nothing: A History of the Noble Gases*, Oxford University Press, 2010

Steven Johnson, *The Invention of Air*, Penguin Books, 2009

Sam Kean, *Caesar's Last Breath: The Epic Story of the Air around Us*, Doubleday, 2017

Nick Lane, *Oxygen: The Molecule That Made the World*, Oxford University Press, 2016

Harold McGee, *Nose Dive: A Field Guide to the World's Smells*, John Murray, 2020

George Monbiot, *Regenesis: Feeding the World without Devouring the Planet*, Penguin Books, 2023

Vaclav Smil, *Enriching the Earth: Fritz Haber, Carl Bosch, and the Transformation of World Food Production*, MIT Press, 2001

Patrick Süskind, *Perfume*, Penguin Books, 2007

Leslie Tomory and Jed Z. Buchwald, *The Origins of the Gaslight Industry, 1780–1820*, MIT Press, 2012

Isabella Tree, *Wilding*, Picador, 2018

Jules Verne, *Around the World in Eighty Days*, Collins Classics, 2018

Matt Winning, *Hot Mess: What on Earth Can We Do about Climate Change?*, Headline, 2023

Acknowledgements

I sincerely thank my editors, Connor Brown and Ivy Givens, for being so patient, supportive and critically incisive.

I work at UCL with a team of scientists, artists, makers, engineers, architects, designers and policy experts. They have all helped me in some way to make this book. I want to thank the whole UCL Institute of Making team, the Plastic Waste Innovation Hub team, the Net Zero Systems team and the UCL Mechanical Engineering Department for their friendship and support.

I would particularly like to thank those who commented on the book as it took shape. Andrea Sella, Sharon Ruston, Ian Anderson, Andy Godfrey, Naomi Gibbs, Daniel Crewe and Buzz Baum (and his Cambridge science writing club) all read chapters of the book and gave me extremely helpful feedback.

I'd like to thank my literary agent, Peter Tallack, who got the book off the ground in the first place, and the whole Penguin Random House team for help with the production process and getting it out into the world. Particular thanks go to Mark Handsley for the copy editing, and to Rosey Battle for organizing copyright permissions for the illustrations.

I am very grateful to Lal Hitchcock, George Wright and Diane Storey, for all the support, and the many restorative Dorset days together, while writing this book.

ACKNOWLEDGEMENTS

I'd like to thank my kids, Lazlo and Ida, for sharing their boundless enthusiasm for gases through COVID lockdowns and helping me with the very entertaining experimental phase of this book.

And finally, I'd like to thank my wonderful wife, Ruby Wright, for all her help with editing and for being my creative inspiration.

Index

Page references in *italics* indicate images.

caliche 238
Cape of Good Hope 141, 145
caramelization 113
carbon
    alcohol and 33
    carbon-based life forms 65
    charcoal and 51
    electric light bulb and 183
    natural gas and 52
    wood fire 50
carbon dioxide xvii, xx, xxi,
    209–26
    breathing and 82–3
    carbon capture 222, 223,
        224, 226
    Carboniferous period and
        66–8, 67
    climate change and xx, xxi,
        209–11, 213, 214, 215–19,
        221–6, 256, 257
    emissions per person 217
    food additives and xxii
    Great Oxygen Event and
        66–7
    greenhouse effect and
        209–10
    molecule, density of 151
    photosynthesis and
        49–50, 253
    sea levels and 209, 211–14, 212
    taxes on 221, 222, 223, 225,
        226, 256
    as waste gas 66
carbon monoxide 26, 54, 61

Carboniferous period 66–8, 67
Carnyx (ancient war trumpets) 85
cars
    air valves and 97–9, 98
    batteries 156
    carbon dioxide emitted from
        210, 225
    catalytic converters 241
    electric 22
    horsepower of 10
    pollution 255
    silicon chips in 187
    steam car 11, 12, 13, 13, 15,
        18, 21–2
    tyres 97, 102
catalysts 241, 242
catalytic converters 241
cathode-ray tubes 185,
    185, 186
Chanel No. 5 124
chaos 146–7
charcoal 51, 52, 55
Cheng Ho (Zheng He) 140, 140
childbirth 36–7, 40–41, 253
China 3, 83, 134, 139, 140–41,
    143, 144, 217
Chincha islands 237, 238
Chinese Zhou Dynasty 3
chisels 86
chlorine gas 229–31, 229
chlorofluorocarbon (CFC)
    gases xviii–xix
chloroform 35–9
Claude, Georges 200

sea levels, climate change and
134, 209, 211–14, *212*, 217,
218, 221, 223, 224, 225–6
Second World War (1939–45)
228, 230
self-balance 96–7
Self-Contained Underwater
Breathing Apparatus
(SCUBA) 63–4, 69–71, *70*,
72, 74, 77, 79
shipbuilding 134–8, *135*, *136*
silicon chips 187
Simpson, Dr John 35–6, 39–40
sliding door, automatic 82, 100
smell xvii
air pollution and 25
ammonia 240
artificial intelligence and 112
'camera' 119
capturing 107–8
'clean' smell 109–11
disorders 120–21
esters and 111, *112*
food and xxii–xxiii, 113,
116–20, 122, 125
fragrance, theory of 123–4
gas lighting and 54–6, 59
labelling experiences and
117–18
liquids and 34, 35
losing sense of 119–21
memory and 106, 116, 119,
125–6
methane 54–5, 56, 59

musk 123–5
patterns in a chemical's
structure associated with
particular odours 111–12,
112
perfume *see* perfume
pheromones 120–23
receptor patterns and 114–15
retronasal 114
temperature and 113–14
soils 117, 223, 231–6, 242, 244,
255
solar power xxi–xxii, *132*
South China Sea 134
Southey, Robert 27
Soviet Union 203–4, *203*
spectral lines 196–7, *197*, 199
spectroscopy 195–7
spirits xvii, xviii, 2, 43–7, 118,
226, 250
steam xvi, xvii, xviii, xxi, xxiii,
4–22, 25, 54, 162
*Around the World in Eighty
Days* and 20–21
automated manufacturing
and 252
boats/ships and 15–17, *16*,
20–21, 100, 167
Boulton and 9, 10, 11, 14, 55,
56
cars and 11, *12*, 13, *13*, 21–2
condensing steam, creation
of large forces through
4–5, *4*